Mechanic's words in five languages

English, Norwegian, Northern Saami, Swedish and Finnish

Petter Reinholdtsen
Oslo

Publisher and technical editor: Petter Reinholdtsen.

Database editor: Svein Lund.

ISBN 978-82-93828-06-8 (hardcover) and 978-82-93828-07-5 (epub).

Dewey 423 439.7 439.8 494.5754 620.003 621.039

http://www.hungry.com/~pere/publisher/.

This book is based on a database with source date from terms collected in the first half of 1990, organized and funded by Samisk videregående skole og reindriftsskole, by editor Svein Lund and several contributors. The same database was the source of the book "Mekanihkkársánit : Mekanikerord = Mekaanisen alan sanasto = Mechanic's words" (ISBN 82-7954-042-3, 1999). Sámi Parliament of Norway and Svein Lund has permitted Petter Reinholdtsen to publish the material with an Creative Commons license CC-BY 4.0.

Buy both paper and electronic edition directly from https://lulu.com/.

https://gitlab.com/petterreinholdtsen/mekanikerord

The book contains about 1200 dictionary entries with translations into other languages. The PDF and paper edition contain a register with words in every language to make it easy to find the relevant dictionary entry.

The words for different languages are marked using language codes according to ISO-639-1: (en) English, (nb) Norwegian Bokmål, (nn) Norwegian Nynorsk, (se) Northern Sami, (sv) Swedish and (fi) Finnish.

The book is based on a Filemaker Pro database with source data from words and terms collected in the first half of the 1990 decade. This work was organized from Samisk videregående skole og reindriftsskole and funded by the same school as well as Samisk utdanningsråd, Nordisk kulturråd and Letterstedtska föreningen. The same database was the source of the book "Mekanihkkársánit : Mekanikerord = Mekaanisen alan sanasto = Mechanic's words" (ISBN 82-7954-042-3, 1999). The editor of the database and book was Svein Lund, who reports that the other contributors were Aimo Aikio, Lars Erik Bønå, Nils Øivind Helander, Klemet I. Hætta, Nils Morten Hætta, Biret Kallio, Øyvind Moeng, Rolf Olsen, Aage Olsen, Anders Oskal, Inger-marie Oskal, Arne Johan Turi, Mai Britt Utsi, Niilo Vuomajoki and Per Aarseth.

Glossary

abrasive cloth, emery cloth

slipelerret (nb) šliipenliidni (se) slipduk, smergelduk (sv) hiomakangas (fi)

abrasive grain, abrasive particle

slipekorn (nb) šliipengordni, šliipenmoallu (se) slipkorn (sv) hiomarae (fi)

abrasive particle

See "abrasive grain, abrasive particle".

absorb

absorbere (nb) absorberet (se) absorbera (sv) absorboida (fi)

accelerator (pedal)

gasspedal (nb) gássaduolmman (se) gaspedal (sv) kaasupoljin (fi)

acceptance limit

gå-grense (nb) čáhkanrádji (se)

accumulator

akkumulator (nb) akkumuláhtor (se) ackumulator (sv) akku, akkumulaattori (fi)

accuracy

See "precision, accuracy".

acetylene

acetylen (nb) acetylena (se) acetylen (sv) asetyleeni (fi)

acid-proof stainless steel

syrefast stål (nb) suvrenanu stálli (se) syrafast stål (sv) haponkestävä teräs (fi)

acute angle

spissvinkel (nb) čohkačiehka (se) spetsvinkel (sv) kärkikulma (fi)

adapter, adaptor

adapter (nb) adapter, lihttu (se) adapter (sv) adapteri, liitin (fi)

adapter sleeve, clamping bush

klemhylse (nb) čárvenskuohppu (se) klämhylsa (sv) puristusholkki (fi)

adaptor

See "adapter, adaptor".

adhesive

See "binding agent, adhesive, agglutinant".

adjust

justere (nb) dárkilastit (se) justera (sv) tarkistaa (fi)

adjustable packing ring cutter

pakningsskjærer[nb], pakningsskjerar[nn] (nb) spintočuohpan, bagadasčuohpan (se) rundskärare (sv) tiiviste leikkuri (fi)

adjustable reamer

stillbar brotsj (nb) molssohahtti galján (se) ställbar brotsch, justerbrotsch (sv) säädettävä kalvain (fi)

adjustable spanner

See "adjustable wrench, adjustable spanner".

adjustable wrench, adjustable spanner

skiftenøkkel (nb) molssačoavdda (se) skiftnyckel (sv) jakoavain (fi)

adjusting, adjustment

justering (nb) dárkilastin (se) justering (sv) tarkistus (fi)

adjusting screw, set screw

justeringsskrue, stillskrue (nb) stellenskruvva (se) justerskruv, ställskruv (sv) säätöruuvi (fi)

adjustment

See "adjusting, adjustment".

admixturing material

See "filler material, admixturing material".

advance

See "feed, advance".

aeration

See "ventilation, venting, airing, aeration".

agglutinant

See "binding agent, adhesive, agglutinant".

aggregate

aggregat (nb) aggregáhta (se) aggregat (sv) agregaatti (fi)

aggregate

(strøm)aggregat, straum-[nn], elektrisk aggregat (nb) rávdnje-aggregáhta (se) aggregat (sv) agrigaatti, sähkögeneraattori (fi)

air bleeding, deaeration, air venting

avlufting (nb) áibmosirren (se) avluftning (sv) ilmanpoisto (fi)

air blowing tube

blåsespiss, luftpistol (nb) bossunnjunni (se) luftpistol (sv) paineilmasuutin, paineilmapistooli (fi)

air cooled

luft(av)kjølt (nb) áibmočoaskuduvvon (se) luftkyld (sv) ilmajäähdytteinen (fi)

air filter

luftfilter (nb) áibmofilttar (se) luftfilter (sv) ilmansuodatin (fi)

air hammer, pneumatic hammer

(trykk)lufthammer[nb], lufthammar[nn], muttertrekker[nb] (nb) áibmoveažir (se) lufthammare, tryckluftshammare (sv) paineilmavasara (fi)

airing

See "ventilation, venting, airing, aeration".

air venting

See "air bleeding, deaeration, air venting".

alignment, straightening

oppretting (nb) njulgen (se) riktning (sv) oikaisu (fi)

allen key

See "hexagon wrench, allen key".

allowance

See "tolerance, allowance, margin".

alloy

legering (nb) metállaseaguhus, seaguhus, segohus (se) legering (sv) seos, metalliseos (fi)

alloy

legere (nb) seaguhit (se) legera (sv) lejeerata, seostaa (fi)

alternate

See "change gear, shift gear, alternate".

alternating current

vekselstrøm, vekselstraum[nn] (nb) molssarávdnji (se) växelström (sv) vaihtovirta (fi)

amplifier

forsterker[nb], forsterkar[nn] (nb) gievrudat (se) förstärkare (sv) vahvistin (fi)

analogic

See "analogous, analogic".

analogous, analogic

analog (nb) analogalaš, analoga (se) analog (sv) analoginen (fi)

angle

vinkel (nb) čiehka (se) vinkel (sv) kulma (fi)

angle

See "bend, elbow, angle, knee".

angle bar

See "angle iron, angle bar, corner iron".

angle gauge

See "square, angle gauge, bevel protractor".

angle gauge, angulometer, bevel protractor

gradvinkel (nb) čiehkamihtádas, grádaviŋkil (se) lutningsmätare, inklinometer (sv) kaltevuusmittari, kulmamittari (fi)

angle iron, angle bar, corner iron

vinkeljern (nb) čiehkaruovdi (se) vinkeljärn (sv) kulmateräs (fi)

angle of thread

flankevinkel, gjengevinkel (nb) gilgačiehka (se) flankvinkel, gängvinkel (sv) kylkikulma (fi)

angle screwdriver, bent screwdriver

vinkelskrujern, vinkeltrekker[nb], vinkeltrekkjar[nn] (nb) čiehkaskruvvaruovdi, roaŋkeskruvvaruovdi (se) vinkelskruvmejsel (sv) kulmaruuvitaltta (fi)

angular grinder

vinkelsliper[nb], vinkelslipar[nn] (nb) čiehkašliipa, viŋkilšliipa (se) vinkelslipmaskin (sv) kulmahiomakone (fi)

angulometer

See "angle gauge, angulometer, bevel protractor".

anneal

anløpe (nb) devkkodahttit (se) anlöpa (sv) päästää (fi)

anneal, glow

gløde (nb) áhcagahttit (se) glödga (sv) hehkuttaa, kumentaa (fi)

annealing

anløping (nb) devkkodahttin (se) anlöpning (sv) päästö, päästäminen (fi)

anti-knock value

bankefasthet[nb], bankefastleik[nn] (nb) dearpannanosvuohta (se) knackningsbeständighet (sv) nakutuksenkestävyys (fi)

anvil, stake

ambolt (nb) stáđđi (se) städ (sv) alasin (fi)

apparatus, device

apparat (nb) apparáhta, rusttet (se) apparat (sv) laite, koje (fi)

arc eye

sveiseblink, sveiseblindhet (nb) sveisensuddon, sveisenšleađggastat (se) svetsblänk (sv) hitsaussokeus, lumisokeus (fi)

armature

See "fittings, armature, fixture".

armouring, reinforcing

armering (nb) nannen, nanosmahtti (se) armering (sv) vahvike (fi)

artisan

See "craftsman, crafter, artisan".

assemble, mount

montere (nb) čohkket, monteret (se) montera (sv) koota (fi)

assembly tong

See "pipe tongs, assembly tong, pipe wrench".

asynchronous

asynkron (nb) asynkruvnnalaš (se) asynkron (sv) asynkroninen, tahdistamaton (fi)

asynchronous motor

asynkronmotor (nb) asynkronmohtor (se) asynkronmotor (sv) epätahtimoottori, asynkroninen moottori (fi)

atomization

See "atomizing, atomization".

atomizing, atomization

forstøvning[nb], forstøving (nb) gavjadeapmi (se) atomisering, finfördelning (sv) sumuttuminen, sumutus (fi)

automatic

automatisk (nb) automáhtalaš (se) automatisk (sv) automaattinen (fi)

automation

See "automatization, automation".

automatization, automation

automatisering, automasjon (nb) automatiseren, automašuvdna (se) automatisering, automation (sv) automaatio (fi)

automaton

automat (nb) automáhta (se) automat (sv) automaatti (fi)

awl

syl (nb) náskál, soairu (se) syl (sv) naskali (fi)

axe

øks (nb) ákšu (se) yxa (sv) kirves (fi)

axial

aksiell (nb) aksiála (se) axial, axiell (sv) aksiaalinen (fi)

axle

See "shaft, axle".

back clearance, end-play angle

endeklaringsvinkel (nb) geahčegaskačiehka (se) baksläppningsvinkel (sv) päästökulma (fi)

back edge of knife

knivrygg, øksehammer[nb], øksehammar[nn] (nb) šimir (se) knivrygg (sv) hamara (fi)

backfire

See "kick, recoil, backfire".

backlash

dødgang, slakk (nb) loaŋkkisteapmi, šloaŋkkisteapmi (se) dödgång, spel (sv) välys, väljyys (fi)

back-rake angle

sponvinkel (nb) vuolahasčiehka (se) spånvinkel (sv) rintakulma (fi)

back-rake surface

sponflate (nb) vuolahasolggoš (se) spånyta (sv) rintapinta, lastupinta (fi)

balance

balanse, likevekt, jamvekt (nb) dássádat (se) balans (sv) tasapaino (fi)

balance, counterbalance

avbalansere, stabilisere (nb) dásset (se) utbalancera,stabilisera (sv) tasapainottaa (fi)

balance, counterweight

motvekt (nb) vuostedeaddu (se) motvikt (sv) vastapaino (fi)

balk

See "beam, balk".

ball

kule (nb) luođđa, jorbadas (se) kula (sv) kuula (fi)

ball bearing

kulelager (nb) luođđaláger (se) kullager (sv) kuulalaakeri (fi)

ball cage, ball retainer

kuleholder[nb], kulehaldar[nn] (nb) luođđadoalan (se) kulhållare (sv) kuulakehikko, kuulanpidin (fi)

ball hammer

kulehammer[nb], kulehammar[nn] (nb) jorbaveažir, duorranveažir (se) kulhammare (sv) kuulapäävasara (fi)

ball retainer

See "ball cage, ball retainer".

banana plug

bananplugg, bananstikker (nb) bananbunci (se) banankontakt (sv) banaanipistoke (fi)

band

See "driving belt, strap, band".

band saw

bandsag (nb) báddesahá (se) bandsåg (sv) vannesaha (fi)

bar

See "handle, lever, bar".

barrel

See "cylinder, roll, roller, barrel".

base

See "foundation, base".

basic

elementær (nb) álggalaš, álgo- (se) elementär (sv) alkio, alkeis- (fi)

battery

batteri (nb) báhtter, batteriija (se) batteri, ackumulator (sv) akku, paristo (fi)

battery charger

batterilader[nb], batteriladar[nn] (nb) báhttergealddán (se) batteriladdare (sv) akkulaturi (fi)

bead

sveisestreng, sveiselarve (nb) sveisensuotna, sveisenstreaŋga (se) sträng (sv) hitsattu palko (fi)

beam, balk

bjelke (nb) bielká (se) balk (sv) palkki (fi)

bearing

lager (nb) láger (se) lager (sv) laakeri (fi)

bearing bush

See "bearing shell, bearing bush".

bearing housing

lagerhus (nb) lágerbeassi (se) lagerhus, lagerbox (sv) laakeripesä (fi)

bearing shell, bearing bush

lagerskål (nb) lágergárri (se) lagerskål (sv) laakerikuori (fi)

become blunt

sløves[nb], sløvast[nn] (nb) nuorrahuvvat, nuorrašuvvat (se) bli slö, bli oskarp (sv) tylsyä (fi)

bed, lathe bed

vange (nb) stivrenbielka (se) prisma (sv) johde, runko, johteet (fi)

beetle

See "mallet, beetle".

bellows

belg, blåsebelg (nb) vuossu (se) bälg (sv) palje (fi)

belt drive

reimdrift (nb) ruopmadoaibma (se) remdrift (sv) hihnakäyttö (fi)

belt pulley

See "pulley, belt pulley".

belt, track

belte (nb) boagán (se) larvband, matta (sv) tela, telamatto (fi)

bench vice, vice

skruestikke (nb) skruvvadoalan, skruvvabasta (se) skruvstycke (sv) ruuvipenkki (fi)

bend, elbow, angle, knee

bend (nb) sodju (se) knärör (sv) käyrä (fi)

bent screwdriver

See "angle screwdriver, bent screwdriver".

benzine

See "gasoline, petrol, benzine".

bevel, chamfer

fas, fase (nb) háloravda, áloravda (se) fas (sv) viiste (fi)

bevel, chamfer

fase (nb) ravdet, ravdadit (se) fasa (sv) viistää (fi)

bevel gear

konisk tannhjul (nb) lávvolat bátnejuvla (se) koniskt kugghjul (sv) kartiohammaspyörä (fi)

bevel protractor

See "angle gauge, angulometer, bevel protractor".

bevel rule, mitre rule

smygvinkel, stillbar vinkel (nb) heivenviŋkil, skivdnjeviŋkil (se) smygvinkel, ställvinkel (sv) säätökulmain, kääntökulmain (fi)

billett saw

See "wood saw, billett saw".

bin

See "container, bin, reservoir, tank".

binary

binær (nb) binára (se) binär (sv) binaarinen (fi)

binding agent, adhesive, agglutinant

bindemiddel (nb) čatnanávnnas (se) bindemedel (sv) sideaine, sidosaine (fi)

bit

bit (nb) bihttá (se) bit (sv) bitti (fi)

blacksmith

See "smith, blacksmith, forger".

blade, scoop, bucket, vane

skovl (nb) soadji, guksi (se) skovel, vinge, blad (sv) siipi (fi)

blast furnace

masovn[nb], masomn[nn] (nb) masomman, masuvdna (se) masugn (sv) masuuni (fi)

block

blokk (nb) blohkka (se) block (sv) lohko, harkko (fi)

block, pulley block

blokk (nb) blohkka (se) block (sv) väkipyörä (fi)

blunt, dull

sløv, skjemt, ukvass (nb) basttoheapmi, ávjjoheapmi (se) slö, oskarp (sv) tylsä (fi)

blunt file

See "flat file, blunt file, safe-edge file".

body

karosseri (nb) goavdi (se) kaross, karosseri (sv) kori (fi)

bolt

bolt (nb) boaltu (se) bult (sv) pultti (fi)

bolt cutter, multipower end cutting nippers

kraftavbiter[nb], kraftavbitar[nn], boltsaks (nb) fápmočuohpan (se) bultsax, kraftavbitare (sv) voimaleikkurit, pulttisakset (fi)

bond

See "earth, ground, bond".

bonding

See "earthing, grounding, bonding".

bore

See "drill, bore".

boring head, drilling head

utboringshode[nb], utboringshovud[nn] (nb) galljudanoaivi (se) arborrhuvud (sv) avarrusistukka, avarruspää (fi)

boss

See "hub, nave, boss".

bottoming tap

See "plug tap, finishing tap, bottoming tap".

bow, clamp, stirrup, loop

bøyle, bøyel (nb) geavli (se) bygel (sv) lenkki, sanka, sinkilä (fi)

bow saw

See "hack saw, bow saw".

box, socket

kopp, pipe (nb) gohppu (se) hylsa (sv) hylsy (fi)

box wrench

See "spanner, box wrench, socket wrench".

box wrench, socket wrench

koppnøkkel, pipenøkkel (nb) gohppočoavdda (se) hylsnyckel (sv) hylsyavain (fi)

brake

bremse (nb) goahcat, vávlet (se) bromsa (sv) jarruttaa (fi)

brake

brems (nb) goazan, vávlu (se) broms (sv) jarru (fi)

brake block

See "brake pad, brake block".

brake disc

bremseskive (nb) goahcanskearru (se) bromsskiva (sv) jarrulevy (fi)

brake drum

bremsetrommel (nb) goahcanfárppal (se) bromstrumma (sv) jarrurumpu (fi)

brake fluid

bremsevæske (nb) goahcannjalbi (se) bromsvätska (sv) jarruneste (fi)

brake pad, brake block

bremsekloss (nb) goahcanbircu, goahcanlohti (se) bromskloss (sv) jarrupala (fi)

brake shoe

bremsesko (nb) goahcangáma (se) bromsback (sv) jarrukenkä (fi)

brass

messing (nb) meisset, messet (se) mässing (sv) messinki (fi)

braze

hardlodde (nb) garrajugahit (se) hårdlöda (sv) kovajuottaa (fi)

brittleness

sprøhet[nb], sprøleik[nn] (nb) smirrodat, smierusvuohta (se) sprödhet (sv) hauraus (fi)

broach

See “ream, broach”.

bronze

bronse (nb) bronsa, bronša (se) brons (sv) pronssi (fi)

brush

børste (nb) gusta, luvda (se) borste (sv) harja, hiiliharja (fi)

bucket

See “blade, scoop, bucket, vane”.

buckrake

See “hay rake, buckrake”.

buffer

See “bumper, buffer”.

built-up edge

løsegg[nb], lausegg (nb) luovusávju (se) lösegg (sv) irtosärmä (fi)

bulb, lamp bulb

lyspære (nb) čuovgapeara, áhcagastinlámpá (se) glödlampa (sv) hehkulamppu (fi)

bumper, buffer

støtfanger[nb], støytfangar[nn] (nb) dustehat (se) stötfångare (sv) puskuri (fi)

burr, rough edge

grad (nb) boazzi (se) grad (sv) purse, jäyste (fi)

bush(ing), sleeve, socket

bøssing, hylse (nb) skuohppu (se) bussning (sv) (laakerin) holkki (fi)

but strap

See “fish joint, junction plate, but strap”.

butt welding

stukesveising (nb) dulpensveisen (se) stuksvetsning (sv) tyssähitsaus (fi)

butt welding bend

sveisebend (nb) sveisenmoalki (se) svetsinsats (sv) kaasuhitsauspilli (fi)

by-pass

omløp (nb) meaddilmannan (se) förbigång, bypass (sv) ohivirtaus (fi)

bypass valve

omløpsventil (nb) meaddilventiila (se) förbigångsventil, bypassventil, shuntvewntil (sv) ohitusventtiili (fi)

by-product

biprodukt, sideprodukt (nb) liigebuvtta (se) biprodukt (sv) sivutuote (fi)

cable clip, cable shoe

kabelsko (nb) jođasgáma (se) kabelsko (sv) johtimen liitin, kaapelikenkä (fi)

cable shoe

See "cable clip, cable shoe".

cable, wire, cord

ledning[nb], leidning[nn], kabel (nb) jođas (se) ledning, kabel (sv) johto, kaapeli (fi)

calibrate

kalibrere (nb) kalibreret (se) kalibrera (sv) kalibroida (fi)

calipers

See "divider, calipers, compasses".

calorific value

See "heating value, calorific value".

camshaft

kamaksel (nb) njunneáksil (se) kamaxel (sv) nokka-akseli (fi)

capacitive

kapasitiv (nb) kapasitiivalaš (se) kapacitiv (sv) kapasitiivinen (fi)

capacitor

See "condenser, capacitor".

capacity

kapasitet (nb) kapasitehta, nákca (se) kapacitet (sv) kapasiteetti (fi)

capillary action

kapillarvirkning[nb], hårrørs-, -verknad[nn] (nb) vuoktabohccedoaibma, kapillardoaibma (se) kapillaritet,hårrörsverkan (sv) hiusputki-ilmiö, kapillaari-ilmiö (fi)

cap, lid, cover

deksel (nb) govččas (se) täckplatta (sv) kansi, konepeitto (fi)

capstan

See "reel, swift, capstan".

carbon, coal

karbon, kull[nb], kol[nn] (nb) čađđa, karbon (se) kol (sv) hiili (fi)

carburettor

forgasser[nb], forgassar[nn] (nb) gássor (se) förgasare (sv) kaasutin (fi)

cardan joint

kardang (nb) kardaŋga (se) kardanknut (sv) nivel (fi)

car, motor-car

bil (nb) biila (se) bil (sv) auto (fi)

carpenter

snekker[nb], snikkar[nn] (nb) snihkkár (se) snickare (sv) puuseppä, kirvesmies (fi)

carpenter

See "joiner, carpenter".

carpenter's bench, planing bench

høvelbenk (nb) heavvalbeaŋka (se) hyvelbänk (sv) höyläpenkki (fi)

carriage

See "wagon, carriage".

carrier, collar

medbringer[nb], medtakar[nn] (nb) jorahanruovdi (se) medbringare (sv) vääntiö, väännin (fi)

carrying capacity, loadbearing capacity

bæreevne[nb], bereevne[nn] (nb) guoddinnákca, guoddináhppi (se) bärförmåga, kapacitet (sv) kantavuus, kuormitettavuus (fi)

case-hardening

settherding (nb) lasihanbuoššodeapmi (se) sätthärdning (sv) hiiletyskarkaisu, pintakarkaisu (fi)

castellated nut

See "castle nut, castellated nut".

caster

See "rowel, pulley, caster".

casting

støpegods[nb], støypegods (nb) leikonávnnas (se) gjutgods (sv) valukappale, valanne (fi)

cast iron

støpejern[nb], støypejer[nn] (nb) leikonruovdi (se) gjutjärn (sv) valurauta (fi)

castle nut, castellated nut

kronemutter (nb) ruvdnomuhtter (se) kronmutter (sv) kruunumutteri (fi)

cast, mould, found

støpe[nb], støype (nb) leiket (se) gjuta (sv) valaa (fi)

catch

See "locking device, catch".

cavitation

kavitasjon (nb) kavitašuvdna (se) kavitation (sv) kavitaatio (fi)

CDI-box

tenningsboks, CDI-boks (nb) cahkkehanboksa, CDI-boksa (se) CDI-laatikko, CDI-rasia (fi)

cement

See "glue, cement".

centre bit, centre drill

senterbor, sentrumsbor (nb) guovddášbovra (se) centrumborr (sv) keskiöpora (fi)

centre drill

See "centre bit, centre drill".

centre punch, punch

kjørner[nb], kjørnar[nn] (nb) merkenluovččan, merkendurra (se) körnare (sv) pistepuikko (fi)

centring gauge

sentrumsvinkel (nb) guovddášviŋkil (se) centrumvinkel (sv) keskiökulma (fi)

chain

kjetting (nb) gehtegat (se) kätting (sv) ketjut (fi)

chain

kjede (nb) viđji, láhkki (se) kedja (sv) ketju (fi)

chain drive

kjededrift (nb) viđjedoaibma (se) kedjedrift (sv) ketjukäyttö (fi)

chain lock

kjedelås (nb) viđjelássa, viđjelohkka (se) kedjelås (sv) ketjulukko (fi)

chain saw

See "power saw, chain saw".

chain wheel

kjedehjul (nb) viđjejuvla (se) kedjehjul (sv) ketjupyörä (fi)

chamfer

See "bevel, chamfer".

change gear, shift gear, alternate

veksle, gire (nb) molsut (se) växla (sv) vaihtaa (fi)

charge

lade (nb) gealdit (se) ladda (sv) ladata, varata (fi)

chaser, chasing tool,threading tool

gjengestål (nb) jeŋgenstálli (se) gängstål (sv) kierreterä, kierteitystyökalu (fi)

chasing tool

See "chaser, chasing tool,threading tool".

check valve, non-return valve

tilbakeslagsventil (nb) máhccanventiila (se) backventil (sv) vastaventtiili, takaiskuventtiili (fi)

chemical combination, chemical compound

kjemisk forbindelse[nb], kjemisk sambinding[nn] (nb) kemiijalaš ovttastus (se) kemisk förening (sv) kemiallinen sidos (fi)

chemical compound

See "chemical combination, chemical compound".

chilling

See "quenching, chilling".

chip

spon (nb) vuolahas, fuolahas, smáhkku (se) spån (sv) lastu (fi)

chip breaker

sponbryter[nb], sponbrytar[nn] (nb) vuolahasdoaján (se) spånbrytare (sv) lastunmurtaja (fi)

chipping, cutting

sponskjærende bearbeiding[nb], sponskjerande til-[nn] (nb) vuolahasčuohppan, vuolahastin (se) spånskärande bearbetning (sv) lastuava työstö (fi)

chipwood

sponplate (nb) vuolahaspláhtta, vuolahasduolbadas (se) spånplatta (sv) lastulevy (fi)

chisel

meisle (nb) luokčat (se) mejsla (sv) taltata (fi)

chisel

meisel (nb) luovččan (se) mejsel (sv) taltta (fi)

choke

struper[nb], strupar[nn], choke (nb) buváhat, šohkka (se) chokespjäll (sv) rikastin (fi)

choke

See "throttle valve, choke".

chrome coat

See "chrome plate, chrome coat".

chrome plate, chrome coat

forkromme (nb) krommadit, krommet (se) förkroma (sv) kromata (fi)

chuck

kjoks, chuck (nb) giddehat (se) chuck (sv) istukka (fi)

chuck wrench

kjoksnøkkel (nb) giddehatčoavdda (se) chucknyckel (sv) istukka-avain (fi)

cinder

See "slag, cinder".

circlip pliers

låseringstang[nb], låseringstong[nn] (nb) lássenrieggabasttat, lohkkadanrieggabasttat (se) fjäderringstång (sv) lukkorengaspihdit (fi)

circuit breaker

See "switch, circuit breaker, interrupter".

circuit diagram

See "wiring diagram, circuit diagram".

circular grinding machine

rundslipemaskin (nb) jorbašliipenmašiidna (se) rundslipmaskin (sv) pyöröhiomakone (fi)

circular pitch, tooth pitch

tanndeling, deling (nb) bátnejuohku (se) kuggdelning, cirkulär delning (sv) hammasjako, jako (fi)

cirkular saw

sirkelsag (nb) gierdosahá (se) cirkelsåg (sv) pyörösaha (fi)

clad welding

påleggssveising (nb) alasveisen (se) påsvetsning (sv) päällehitsaus (fi)

clamp

See "bow, clamp, stirrup, loop".

clamp arbor

oppspenningsdor (nb) giddenáksil (se) spänndorn (sv) kiinnityskara, kiinnitystuurna (fi)

clamp, clamping plate

klemjern (nb) čárvenruovdi (se) spännklov (sv) puristusrauta (fi)

clamp frame, screw clamp

skrutvinge (nb) skruvvačárvvon, lávgenruovdi (se) skruvtving (sv) ruuvipuristin (fi)

clamping bush

See "adapter sleeve, clamping bush".

clamping jaw

spennbakke (nb) čavgenoalul (se) spännback (sv) kiristysleuka (fi)

clamping plate

See "clamp, clamping plate".

claw coupling

klokopling (nb) gazzalakta (se) klokoppling (sv) kynsikytkin (fi)

clean

See "polish, finish, dress, clean".

clean off burrs, remove burrs

grade (nb) boazzuhuhttit (se) grada (sv) poistaa jäysteet (fi)

clearance angle, relief angle

klaringsvinkel, frivinkel (nb) gaskačiehka (se) släppningsvinkel (sv) päästökulma (fi)

clearance surface, flank

klaringsflate, friflate (nb) gaskaolggoš (se) släppningsyta (sv) päästöpinta (fi)

climb milling, down milling

medfresing (nb) miehttejursan (se) medfräsning (sv) myötäjyrsintä (fi)

clinch

See "close a rivet, upset, clinch, jalt".

close a rivet, upset, clinch, jalt

stuke (nb) dulpet (se) stuka (sv) tyssätä (fi)

closed circuit

slutta strømkrets[nb], lukka krets[nb], slutta krins[nn], lukka krins[nn] (nb) giddejuvvon rávdnjebiire (se) sluten stömkrets (sv) suljettu piiri (fi)

clutch

kløtsj, kopling (nb) lavtta (se) koppling (sv) kytkin, liitin (fi)

coal

See "carbon, coal".

coarse file

grovfil (nb) roavvafiilu (se) grovfil (sv) karkeaviila, rouhintaviila (fi)

coarse file, rasp

rasp (nb) ráspa (se) rasp (sv) raspi (fi)

coarse-grained

grovkorna (nb) roavvagortnat (se) grovkornig (sv) karkearakeinen, - jyväinen (fi)

coat, coating, facing, lining, surfacing

belegg (nb) gokčangeardi (se) beläggning (sv) pinnoite (fi)

coating

See "coat, coating, facing, lining, surfacing".

cock, tap, faucet

kran (nb) hátna (se) kran (sv) hana (fi)

coefficient of linear thermal expansion

lengdeutvidingskoeffisient (nb) guhkidangorri (se) längdutvidgningskoefficient (sv) pituuslaajenemiskerroin (fi)

cogwheel

See "gear wheel, cogwheel, toothed wheel".

coil

See "wind, coil, spool, twist".

coiling

See "winding, coiling".

coil, spool, reel

spole (nb) giesttus (se) spole (sv) kela, käämi (fi)

coke

koks (nb) koksa (se) koks (sv) koksi (fi)

cold saw

kaldsag (nb) čoaskasahá (se) kallsåg (sv) kylmäsaha (fi)

collar

See "carrier, collar".

combination pliers

kombinasjonstang[nb], universaltang[nb], -tong[nn] (nb) lotnolasbasttat (se) kombinationstång (sv) yleispihdit (fi)

combination wrench

stjernefastnøkkel, kombinasjonsnøkkel (nb) lotnolasčoavdda, násterabasčoavdda (se) u-ringnyckel (sv) kiintolenkkiavain (fi)

combustible, inflammable

brennbar (nb) buolihahtti, cahkehahtti (se) brännbar (sv) palava, syttyvä (fi)

combustion

forbrenning (nb) boaldin (se) forbränning (sv) polttaminen (fi)

combustion engine

forbrenningsmotor (nb) boaldinmohtor (se) förbränningsmotor (sv) polttomoottori (fi)

company

See "manufacturing plant, works, company".

compasses

See "divider, calipers, compasses".

compass saw

See "tenon saw, compass saw, dowel saw".

compensate

kompensere (nb) buhtadit (se) kompensera, utjämna (sv) hyvittää, korvata, tasata, kompensoida (fi)

compress

komprimere (nb) čárvet, komprimeret (se) komprimera (sv) puristaa kokoon (fi)

compressed air

trykkluft (nb) deattaáibmu (se) tryckluft (sv) paineilma (fi)

compression

kompresjon (nb) komprešuvdna, čárva (se) kompression (sv) puristus (fi)

compression gauge, compression tester

kompresjonsmåler[nb], kompresjonsmålar[nn] (nb) komprešuvdnamihtádas, čárvamihtádas (se) kompressionsmätare (sv) puristuspainemittari (fi)

compression tester

See "compression gauge, compression tester".

compressor

kompressor (nb) kompressor (se) kompressor (sv) kompressori (fi)

computer

datamaskin (nb) dihtor (se) dator, datamaskin (sv) tietokone (fi)

concentrate

See "enrich, dress, concentrate".

concentration

See "enrichment, dressing, concentration".

concrete

betong (nb) betoŋga (se) betong (sv) betoni (fi)

condensate

kondens (nb) kondeansa (se) kondens (sv) kondenssi, lauhde (fi)

condensation

kondensering (nb) kondenseren (se) kondensering (sv) kondensointi, lauhtuminen (fi)

condensed water remover

kondensfjerner[nb], kondensfjernar[nn] (nb) kondeansajávkkadas, denna (se) karburatorsprit (sv) (polttoaineen) jäänestoaine (fi)

condenser, capacitor

kondensator (nb) kondensáhtor (se) kondensator (sv) kondensaattori (fi)

conducting material

See "conductor, conducting material".

conductor, conducting material

leder[nb], leiar[nn] (nb) jođadas, jođasávnnas (se) ledare (sv) johde (fi)

cone

See "tang, cone".

conebelt, V-belt

kilreim, kilereim (nb) lohteruopma (se) kilrem (sv) kiilahihna (fi)

coned

See "conical, coned, tapered, coniform".

conical, coned, tapered, coniform

konisk (nb) lávvolat, čohkolaš (se) konisk (sv) kartiokas (fi)

coniform

See "conical, coned, tapered, coniform".

connecting rod

See "piston rod, connecting rod".

construction

bygg (nb) huksehus (se) bygge, byggnad (sv) rakennus (fi)

construction

See "design, construction".

construction machine

anleggsmaskin (nb) ráhkadusmašiidna (se) anläggningsmaskin (sv) työkone (fi)

construction steel, structural steel

konstruksjonsstål (nb) ráhkadusstálli (se) konstruktionsstål (sv) rakenneteräs (fi)

construction works

See "plant, construction works".

consumption

forbruk (nb) gollu, mannu (se) förbrukning (sv) käyttö (fi)

contact

See "socket, contact".

contact breaker

See "contact set, contact breaker".

contact face, sealing face

tetningsflate[nb], tettingsflate (nb) lávgehat (se) tätningsyta (sv) tiivistyspinta (fi)

contact gap

kontaktåpning[nb], -opning[nn], kontaktavstand (nb) oktavuođagoavki (se) kontaktavstånd (sv) katkojan kärkiväli (fi)

contactor

kontaktor (nb) kontaktor (se) kontaktor (sv) kontaktori (fi)

contact point

målespiss (nb) mihtidannjunni (se) mätspets (sv) mittakärki (fi)

contact set, contact breaker

platinastift, stift, avbryterkontakt[nb], avbrytar-[nn] (nb) dollaveažir (se) brytarspets (sv) katkojan kärki (fi)

container, bin, reservoir, tank

beholder[nb], behaldar[nn], tank (nb) tánka, lihtti, stámpa (se) tank, behållare (sv) säiliö (fi)

control

kontrollere (nb) dárkkistit (se) kontrollera (sv) tarkastaa (fi)

control

See "inspection, control".

control engineering

reguleringsteknikk (nb) reglerenteknihkka (se) reglerteknik (sv) säätötekniikka (fi)

control technique

styringsteknikk (nb) stivrenteknihkka (se) styrteknik (sv) ohjaustekniikka (fi)

control unit

styreenhet[nb], styreeining[nn] (nb) stivrran (se) styrenhet (sv) ohjain, ohjausosa (fi)

conventional milling, up milling

motfresing (nb) vuostejursan (se) motfräsning (sv) vastajyrsintä (fi)

converter

konverter (nb) konverter (se) konverter (sv) konvertteri (fi)

converter

omformer[nb], omformar[nn] (nb) muhtán (se) omformare (sv) muuntaja (fi)

conveyor, transporter

transportør (nb) fievrran (se) transportör (sv) kuljetin (fi)

coolant

kjølevæske (nb) čoaskudannjalbi (se) kylvätska (sv) jäähdytysneste (fi)

cooling system fitter

kuldemontør (nb) galmmihanmekanihkkár (se) kylmontör (sv) kylmäkonekorjaaja, kylmäkoneasentaja (fi)

coordinate

koordinat (nb) koordináhta (se) koordinat (sv) koordinaatti (fi)

copper

kopper[nb], kopar[nn] (nb) veaiki (se) koppar (sv) kupari (fi)

cord

See "cable, wire, cord".

corner iron

See "angle iron, angle bar, corner iron".

corner radius

neseradius (nb) njunneradius (se) nosradie (sv)

corrode

See "rust, corrode".

corrosion

korrosjon (nb) korrošuvdna (se) korrosion (sv) korroosio, syöpyminen (fi)

cotter

See "splint, splinter, pin, cotter, split pin".

counterbalance

See "balance, counterbalance".

counter nut, locknut

kontramutter (nb) vuostemuhtter (se) kontramutter (sv) vastamutteri (fi)

countershaft, intermediate shaft

mellomaksel (nb) gaskaáksil (se) mellanaxel (sv) väliakseli (fi)

countersink

forsenker[nb], forsenkar[nn], forsenkingsbor (nb) goban (se) försänkare (sv) upotuspora (fi)

countersink

forsenke (nb) gohpat (se) försänka (sv) upottaa (fi)

countersunk head

forsenka hode[nb], forsenka hovud[nn], senkhode[nb], senkhovud[nn] (nb) gohpeoaivi (se) försänkt huvud (sv) uppokanta (fi)

counterweight

See "balance, counterweight".

couple

kople (nb) laktit, goallustit (se) koppla (sv) kytkeä, liittää (fi)

coupling

kopling (nb) lavtta, goallus (se) koppling (sv) liitin, kytkin (fi)

cover

See "cap, lid, cover".

crafter

See "craftsman, crafter, artisan".

craftsman, crafter, artisan

håndtverker[nb], handverkar[nn] (nb) duojár (se) hantverkare (sv) käsityöläinen (fi)

crane

kran, heisekran (nb) lovtton, vinta (se) lyftkran (sv) nosturi (fi)

crane carriage

See "travelling trolley, crane carriage".

crankcase

veivhus, veivkasse (nb) veiveviessu, veivekássa (se) vevhus (sv) kampikammio (fi)

crank, handle

sveiv, veiv (nb) veive, veaiva (se) vev (sv) kampi, veivi (fi)

crank pin

veivtapp (nb) veiveculci (se) vevtapp (sv) kammentappi (fi)

crankshaft

veivaksel (nb) veiveáksil (se) vevaxel (sv) kampiakseli (fi)

crimp

See "shrink on, crimp".

cross bar

See “traverse, transom, cross beam, cross bar”.

cross beam

See “traverse, transom, cross beam, cross bar”.

cross chisel

kryssmeisel (nb) ruossaluovččan (se) kryssmejsel (sv) ristitaltta (fi)

cross edge

See “lateral edge, cross edge”.

cross slide, cross tool slide

tverrsleide (nb) doaresmielggas (se) tvärslid (sv) poikittaisluisti, poikkitaiskelkka (fi)

cross-slot screwdriver

stjerneskrujern (nb) násteskruvvaruovdi (se) krysskruvmejsel (sv) ristikärkitaltta (fi)

cross tool slide

See “cross slide, cross tool slide”.

crow bar

brekkjern, kubein (nb) gákkan, ruovdegákkan (se) bräckjärn, kofot (sv) sorkkarauta (fi)

crow bar, lever, iron bar

spett (nb) ruovdegáŋga (se) spett (sv) rautakanki (fi)

crude oil

råolje (nb) álgguviđá olju, luondduviđá olju (se) råolja (sv) raakaöljy (fi)

current

strøm, straum[nn] (nb) rávdnji (se) ström (sv) virta (fi)

cursor

markør (nb) seaván (se) markör (sv) kohdistin (fi)

cushion, damp

dempe (nb) váidudit (se) dämpa (sv) vaimentaa,hiljentää (fi)

cushioning

demping (nb) váidudeapmi (se) dämpning (sv) vaimennin (fi)

cut, notch, section

tverrsnitt, snitt (nb) sneaktačuohpastat, čuohpastat, sneaktagovva (se) snitt (sv) leikkaus, poikkileikkaus (fi)

cut off

stikke av (nb) čugget, gaskatčuohppat (se) sticka av (sv) katkaista, pistää (fi)

cut threads

See “thread, cut threads”.

cutting

See “chipping, cutting”.

cutting angle, edge angle

eggvinkel (nb) ávjočiehka (se) eggvinkel (sv) teräkulma (fi)

cutting burner

See "cutting torch, cutting burner".

cutting depth

kuttdybde[nb], kuttdjupn[nn] (nb) čuohppančikŋodat (se) skärdjup (sv) leikkuusyvyys, lastuamissyvyys (fi)

cutting die, threading die, screw die

gjengebakke, gjengesnitt (nb) jeŋgenbáhkka (se) gängback, gängsnitt (sv) kierreleuka, kierrepakka (fi)

cutting disc

kutteskive (nb) čuohppanskearru (se) kapslipskiva (sv) katkaisulaikka (fi)

cutting edge

See "edge, cutting edge".

cutting edge angle

See "setting angle, cutting edge angle".

cutting off

avstikking (nb) čuggen, gaskatčuohppan (se) avstickning (sv) katkaisu, leikkuu, leikkaaminen (fi)

cutting pliers

avbitertang[nb], avbitartong[nn] (nb) cikcenbasttat, botkenbasttat (se) avbitartång (sv) leikkurit, leikkuripihdit (fi)

cutting speed

skjærehastighet[nb], skjerefart[nn] (nb) čuohppanleaktu (se) skärhastighet (sv) leikkuunopeus, lastuamisnopeus (fi)

cutting torch, cutting burner

skjærebrenner[nb], skjerebrennar[nn] (nb) dollačuohppan (se) skärbrännare (sv) leikkauspoltin (fi)

cycle, period

periode, syklus (nb) birrajohtu (se) period (sv) jakso (fi)

cyclic

See "periodical, cyclic, intermittent, pulsating".

cylinder

sylinder (nb) sylinddar (se) cylinder (sv) sylinteri (fi)

cylinder block, engine block

motorblokk, sylinderblokk (nb) mohtorgorut, sylinddargorut (se) cylinderblock (sv) sylinteriryhmä (fi)

cylinder cover

See "cylinder head cover, cylinder cover".

cylinder head cover, cylinder cover

topplokk (nb) gieralohkki, bajildaslohkki (se) cylinderhuvud, topplock (sv) sylinterinkansi (fi)

cylinder leak tester

sylinderlekkasjetester[nb], -testar[nn] (nb) sylinddarsuođđaniskkan (se) vuotomittari (fi)

cylinder, roll, roller, barrel

vals, valse (nb) ludni, válsa (se) vals (sv) valssi (fi)

cylindrical cutter, plain milling cutter

valsfres (nb) válsajurssan (se) valsfräs (sv) lieriöjyrsin (fi)

damp

See "cushion, damp".

damper

See "flap, damper, valve disc".

data

data (nb) dáhta, atta (se) data (sv) tieto, data (fi)

dead centre (lower / top)

dødpunkt (nedre / øvre) (nb) jápmačuokkis, jorggihansadji (bajit / vuolit) (se) dödpunkt (undre / övre) (sv) kuolokohta (ala-/ylä-) (fi)

dead centre peak

See "dead centre point, dead centre peak".

dead centre point, dead centre peak

fast senterspiss (nb) gitta guovddášnjunni (se) fast dubb (sv) kiinteä keskiökärki (fi)

deaeration

See "air bleeding, deaeration, air venting".

deform

deformere (nb) deahpanit (se) deformera (sv) muotava (fi)

deformation

formendring (nb) hápmemuhttin, lávvamuhttin (se) formförändring (sv) muodonmuutos (fi)

densimeter

kjølesystemtester[nb], kjølesystemtestar[nn] (nb) čoaskudanrusttetiskkan (se) täthetsmätare (sv) jäähdyttimen koestuslaite (fi)

density

tetthet[nb], tettleik[nn], densitet, massetetthet (nb) čoahkkisvuohta (se) densitet (sv) tiheys (fi)

depression

See "low pressure, negative pressure, depression".

design, construction

konstruksjon (nb) ráhkadus, konstrukšuvdna (se) konstruktion (sv) rakenne, konstruktio (fi)

detector, sensor, probe

føler[nb], følar[nn], sensor, giver[nb], gjevar[nn] (nb) dovddan, attán (se) givare, sensor (sv) ilmaisin, anturi, sensori (fi)

deviation, divergence

avvik (nb) gáidu (se) avvikelse (sv) poikkema (fi)

device

See "apparatus, device".

dial gauge, dial indicator

måleur (nb) mihtidandiibmu (se) mätklocka (sv) mittakello (fi)

dial indicator

See "dial gauge, dial indicator".

diaphragm

See "membrane, diaphragm".

die

See "jaw, die".

diesel engine

dieselmotor (nb) dieselmohtor (se) dieselmotor (sv) dieselmoottori (fi)

diesel oil

dieselolje, diesel, solar (nb) dieselolju, diesel (se) dieselolja (sv) dieselöljy (fi)

die stock

svingjern (nb) gávvaruovdi, jorreruovdi (se) svängjärn, gängkloppa (sv) väännin, kierresorkka (fi)

differential

differensial (nb) differensiála (se) differential (sv) tasauspyörästö (fi)

differential lock

differensialsperre (nb) differensiálgiddehat (se) differentialspärr (sv) tasauspyörastön lukitus (fi)

digital

digital (nb) digitálalaš, digitála (se) digital (sv) digitaalinen, numeerinen (fi)

dimension, measure out

dimensjonere (nb) mihttodallat (se) dimensionera (sv) mitoittaa (fi)

diode

diode (nb) dioda (se) diod (sv) diodi (fi)

direct current

likestrøm, likestraum[nn] (nb) dásserávdnji (se) likström (sv) tasavirta (fi)

direction

retning (nb) hálti, guovlu (se) riktning (sv) suunta (fi)

directional valve

retningsventil (nb) hálteventiila (se) riktningsventil (sv) suuntaventtiili (fi)

direction of current, flow direction

strømretning, straumretning[nn] (nb) rávdnjehálti (se) strömriktning (sv) virransuunta (fi)

directory

katalog (nb) ohcu (se) katalog, bibliotek (sv) hakemisto (fi)

disc brake

skivebrems (nb) skearrogoazan (se) skivbroms (sv) levyjarru (fi)

disc, plate, washer

skive (nb) skearru, gearru (se) skiva (sv) levy (fi)

dismantle

See "dismount, dismantle".

dismount, dismantle

demontere (nb) burgit (se) demontera (sv) purkaa (fi)

display

See "screen, monitor, display".

distributor

fordeler[nb], fordelar[nn], strømfordeler[nb], straum-[nn] (nb) juogan, rávdnjejuogan (se) fördelare (sv) virranjakaja (fi)

divergence

See "deviation, divergence".

divider, calipers, compasses

passer[nb], passar[nn], stikkpasser (nb) gierdodahkki (se) passare (sv) harppi (fi)

dividing head

delehode[nb], delehovud[nn] (nb) juohkinoaivi (se) delningsdocka, delningshuvud (sv) jakopää (fi)

dividing plate

See "index plate, dividing plate".

double acting cylinder

dobbeltvirkende sylinder[nb], dobbeltverkande sylinder[nn] (nb) guovttedoaimmat sylinddar (se) dubbelverkande cylinder (sv) kaksitoiminen sylinteri (fi)

double row bearing

See "dual bearing, double row bearing".

dowel

See "pin, dowel".

dowel saw

See "tenon saw, compass saw, dowel saw".

down milling

See "climb milling, down milling".

drain

See "outlet, drain".

drain pipe, outlet tube, sewer pipe

avløpsrør, kloakkrør (nb) duolvačáhcebohcci (se) avloppsrör (sv) viemäriputki (fi)

drain plug

avtappingsplugg, dreneringsplugg (nb) gurrenculci (se) avtappningsplugg (sv) tyhjennystulppa (fi)

drawing

tegning[nb], teikning[nn] (nb) sárggus (se) ritning (sv) piirustus (fi)

drawing board

tegnebrett[nb], teiknebrett[nn] (nb) sárgunbreahtta (se) ritbräde (sv) piirustuslauta (fi)

dredger

See "excavator, dredger".

dress

See "enrich, dress, concentrate".

dressed ore, ore concentrate

slig (nb) málbmariggudus, riggudus (se) slig (sv) malmirikaste, rikaste (fi)

dresser

See "planing tool, dresser".

dressing

See "enrichment, dressing, concentration".

drift, punch

dor (nb) durra (se) dorn (sv) tuurna (fi)

drill

bor (nb) bovra, nábár, rádna, málgur (se) borr (sv) pora (fi)

drill, bore

bore (nb) bovret, bohkat, rádnet (se) borra (sv) porata (fi)

drill coupling

See "tapered sleeve, drill coupling".

drill, drilling machine

boremaskin (nb) bovrenmašiidna, rádna (se) bormaskin (sv) porakone (fi)

drilling head

See "boring head, drilling head".

drilling machine

See "drill, drilling machine".

drive

drivverk (nb) jođihandoaimmahat, jođihanrusttet (se) drivverk (sv) käyttölaite, käyttömekanismi (fi)

drive fit

See "tight fit, drive fit".

driving belt

See "transmission belt, driving belt".

driving belt, strap, band

reim (nb) ruopma, lávži (se) rem (sv) hihna (fi)

driving gear

See "driving wheel, driving gear, driving pinion".

driving pinion

See "driving wheel, driving gear, driving pinion".

driving wheel, driving gear, driving pinion

drev, drivhjul (nb) jođihanjuvla (se) drev, drivhjul (sv) käyttöpyörä, vetopyörä (fi)

drum, barrel

trommel (nb) fárppal (se) trumma (sv) rumpu (fi)

drum brake

trommelbrems (nb) fárppalgoazan (se) trumbroms (sv) rumpujarru (fi)

dual bearing, double row bearing

torada lager (nb) guovtteráiddoláger (se) tvåradigt lager (sv) kaksirivinen laakeri (fi)

ductile

See "forgeable, malleable, ductile".

ductile, plastic

plastisk, formbar (nb) plástalaš, hábmehahtti (se) plastisk (sv) muovailtava, muovautuva (fi)

dull

See "blunt, dull".

dustbin

See "garbage can, trash barrel, dustbin".

dust cleaner

See "vacuum cleaner, dust cleaner".

dynamical

See "dynamic, dynamical".

dynamic, dynamical

dynamisk (nb) dynámalaš (se) dynamisk (sv) dynaaminen (fi)

dynamo

See "generator, dynamo".

earmuff

øreklokker, øyreklokker[nn] (nb) bealljegohput (se) hörselskyddskåpa (sv) korvakupit, korvakuvut (fi)

ear plug

ørepropp, øreplugg, øyre-[nn] (nb) bealljebunci (se) hörselskyddspropp, öronpropp (sv) korvatulppa (fi)

ear protection, hearing protection

hørselsvern (nb) bealljesuojan, gullosuojan (se) hörselskydd (sv) kuulonsuojain (fi)

earth conductor

See "earth wire, earth lead, earth conductor".

earth, ground, bond

jorde (nb) ednet (se) jorda (sv) maadoittaa (fi)

earthing, grounding, bonding

jording (nb) ednen (se) jordning (sv) maadoitus (fi)

earth lead

See “earth wire, earth lead, earth conductor”.

earth wire, earth lead, earth conductor

jordledning[nb], jordingsledning[nb], jordleidning[nn], jordingsleidning[nn] (nb) ednenjođas (se) jordledare (sv) maajohdin, maadoitusjohdin ukkosenjohdin (fi)

eccentricity, run-out

kast, eksentrisitet (nb) skieivun (se) kast (sv) heitto (fi)

edge angle

See “cutting angle, edge angle”.

edge, cutting edge

egg, skjær[nb], skjer[nn] (nb) ávju (se) egg (sv) terä, leikkuusärmä (fi)

edge raising

See “flanging, edge raising”.

effect

effekt (nb) beaktu (se) effekt (sv) teho (fi)

efficiency

virkningsgrad[nb], verknadsgrad[nn] (nb) váikkuhusmearri, doaibmanmearri (se) verkningsgrad (sv) hyötysuhde (fi)

ejector

utslager[nb], utslagar[nn], borkile (nb) luvvenlohti (se) utstötare (sv) irrotin, ulostyönnin, työntötuurna (fi)

elastic, flexible

elastisk (nb) fadnil, vadnil, dávggas (se) elastisk (sv) joustava (fi)

elbow

See “bend, elbow, angle, knee”.

elbow bend

See “elbow, elbow bend”.

elbow, elbow bend

albue[nb], olboge[nn], rørkne (nb) gardnjil (se) rörkrök (sv) putken mutka (fi)

electric(al)

elektrisk (nb) elektralaš, elektrihkalaš (se) elektrisk (sv) sähköinen (fi)

electric arc

lysbue[nb], lysboge[nn] (nb) čuovgadávgi (se) ljusbåge (sv) valokaari (fi)

electric circuit

strømkrets[nb], straumkrins[nn] (nb) rávdnjebiire (se) strömkrets (sv) virtapiiri (fi)

electrician

elektriker[nb], elektrikar[nn] (nb) elektrihkkár (se) elektriker (sv) sähköasentaja (fi)

electricity

elektrisitet (nb) elektrisitehta (se) elektricitet (sv) sähkö (fi)

electrode

elektrode (nb) elektroda (se) elektrod (sv) elektrodi (fi)

electronic

elektronisk (nb) elektrovnnalaš (se) elektronisk (sv) sähköinen, elektroninen (fi)

electronics

elektronikk (nb) elektronihkka (se) elektronik (sv) elektroniikka (fi)

elementary charge

elementærladning (nb) álgogealdu, vuođđogealdu (se) elementärladdning (sv) alkeisvaraus (fi)

emergency stop device, e. shutdown

nødstopp, naudstopp[nn] (nb) heahteorustahtti (se) nödstopp (sv) hätäpysäytin (fi)

emery cloth

See "abrasive cloth, emery cloth".

emery paper, sandpaper

sandpapir (nb) sáttobábir, šliipenbábir (se) sandpapper (sv) hiekkopaperi (fi)

emulsion, solution

emulsjon, oppløsning[nb], oppløysing (nb) luvvadus (se) emulsion, lösning (sv) emulsio (fi)

end face mill

See "shell end mill, end face mill".

end mill

pinnefres (nb) sággejurssan (se) pinnfräs (sv) tappijyrsin, varsijyrsin (fi)

end-play angle

See "back clearance, end-play angle".

energy

energi (nb) árja, energiija (se) energi (sv) energia (fi)

engagement, mesh

inngrep (nb) roahkkohus (se) ingrepp (sv) kosketus (fi)

engine

See "motor, engine".

engine anchorage, engine bracket, engine lug

motorfeste (nb) mohtorgiddehat (se) motorfäste (sv) moottorin kiinnitin (fi)

engine block

See "cylinder block, engine block".

engine bracket

See "engine anchorage, engine bracket, engine lug".

engine heater

motorvarmer[nb], motorvarmar[nn] (nb) mohtorliekkan (se) motorvärmare (sv) moottorilämmitin (fi)

engine lug

See "engine anchorage, engine bracket, engine lug".

enrich, dress, concentrate

anrike[nb], opprike[nn] (nb) riggudit (se) anrika (sv) rikastaa (fi)

enrichment, dressing, concentration

anriking[nb], oppriking[nn] (nb) riggudeapmi (se) anrikning (sv) rikastaminen, rikastus (fi)

e. shutdown

See "emergency stop device, e. shutdown".

etch

etse (nb) borrat (se) etsa (sv) syövyttää, etsata (fi)

excavator, dredger

gravemaskin (nb) goaivunmašiidna (se) grävmaskin (sv) kaivinkone (fi)

excentric

eksentrisk (nb) eahpeguovddášlaš (se) excentrisk (sv) epäkeskinen (fi)

exchange

See "gearing, transmission, exchange".

exhaust

eksos, avgass (nb) bázahasgássa, eksosa, eksosgássa (se) avgas (sv) pakokaasu (fi)

exhaust analyzer

avgassanalysator, avgassmåler[nb], eksosmåler[nb], avgassmålar[nn], eksosmålar[nn] (nb) bázahasmihtádas, eksosmihtádas (se) avgasmätare (sv) pakokaasumittari (fi)

exhaust box

See "silencer, damper, muffler, exhaust box".

exhaust pipe

eksosrør (nb) bázahasbohcci, eksosbohcci (se) avgasrör (sv) pakoputki (fi)

exhaust system

eksosanlegg, avgassanlegg (nb) bázahasrusttet (se) avgassystem (sv) pakoputkisto (fi)

expansion

ekspansjon (nb) báisan (se) expansion (sv) laajeneminen, leviäminen, kasvu (fi)

expansion bolt, expansion-shell bolt

ekspansjonsbolt (nb) báisaboaltu (se) expanderbult (sv) kiilapultti, paisuntakuoripultti (fi)

expansion bush, expansion sleeve

ekspansjonshylse (nb) báisaskuohppu (se) expansionshylsa, expanderhylsa (sv) kiristysholkki, kiilaholkki (fi)

expansion-shell bolt

See "expansion bolt, expansion-shell bolt".

expansion sleeve

See "expansion bush, expansion sleeve".

explosion

eksplosjon (nb) bávkkiheapmi, bávkkehus (se) explosion (sv) räjähdys (fi)

extension bar, extension rod

forlenger[nb], forlengjar[nn] (nb) joatkkán, joatkka (se) förlängningsstång (sv) jatke, jatkovarsi (fi)

extension cord

skjøteledning[nb], skøyteleidning[nn] (nb) joatkkajođas, goallusjođas (se) förlängningssladd, skarvsladd (sv) jatkojohto (fi)

extension rod

See "extension bar, extension rod".

eye bolt

See "eyebolt, eye bolt".

eyebolt, eye bolt

øyebolt[nb], augebolt[nn] (nb) čalbmeboaltu (se) ögleskruv, länkskruv, lyftögla (sv) silmukkaruuvi, nostosilmukka (fi)

face

See "plane, face".

face chuck

See "face plate, face chuck".

face, dress

plandreie (nb) duolbbasinjorahit (se) plansvarva (sv) tasosorvata (fi)

face plate, face chuck

planskive (nb) duolbaskearru (se) planskiva (sv) tasolaikka (fi)

facing

See "coat, coating, facing, lining, surfacing".

facing cutter, surface cutter

planfres (nb) dulbenjurssán (se) planfräs (sv) tasojyrsin, otsajyrsin (fi)

factory, manufacturing plant

fabrikk (nb) fabrihkka (se) fabrik (sv) tehdas (fi)

fan

vifte (nb) boson, vállu (se) fläkt (sv) puhallin, tuuletin (fi)

fan belt

viftereim (nb) bosonruopma, vālloruopma (se) fläktrem (sv) tuulettimen hihna (fi)

fasten, tighten, fix

spenne opp, spenne fast, feste (nb) giddet (se) spänna fast, fästa (sv) kiinnittää (fi)

fatigue

utmatting (nb) váibadahttin (se) utmattning (sv) väsyminen (fi)

faucet

See "cock, tap, faucet".

fault finding

See "fault location, fault finding".

fault location, fault finding

feilsøking (nb) vihkeohcan (se) felsökning (sv) vian paikannus (fi)

feed

mate (nb) borahit (se) mata (sv) syöttää (fi)

feed, advance

mating (nb) boraheapmi (se) matning (sv) syöttö (fi)

feed pump

matepumpe (nb) borahanbumpa, seavdinbumpa (se) matarpump (sv) syöttöpumppu (fi)

feed rod

mateaksel (nb) borahanáksil (se) matarspindel (sv) syöttöakseli (fi)

feed worm, screw feeder

mateskrue, skruetransportør (nb) borahanskruvva (se) skruvtransportör (sv) syöttöruuvi, ruuvikuljetin (fi)

feeler gauge

bladsøker[nb], søkjar[nn] (nb) goavkedulka (se) bladmått, slitsmått (sv) rakomitta, rakotulkki (fi)

felloe

See "rim, felly, felloe".

felly

See "rim, felly, felloe".

felt

filt (nb) duohppa (se) filt (sv) huopa (fi)

fertilizer distributor

gjødselspreder[nb], gjødselspreiar[nn] (nb) muhkebiđggon (se) gödselspridare (sv) lannanlevittäjä, -levitin (fi)

field

felt (nb) gieddi (se) fält (sv) kenttä (fi)

filament

glødetråd (nb) áhcagastinárpu (se) glödtråd (sv) hehkulanka (fi)

file

file (nb) fiilet (se) fila (sv) viilata (fi)

file

fil (nb) fiilu (se) fil (sv) viila (fi)

file

fil (nb) vuorká, fiila (se) fil (sv) tiedosto (fi)

file brush

filbørste (nb) fiilogušta (se) filborste (sv) viilaharja (fi)

filing vice, hand vice

filklo (nb) fiilogazza (se) filklove (sv) viilain, ruuvipuristin (fi)

fill

See "putty, fill".

filler material

See "fillings, filler material".

filler material, admixturing material

tilsettmateriale, tilsett, tilsatsmateriale (nb) ligeávnnas (se) tillsatsmaterial (sv) lisäaine (fi)

fillet weld

kilsveis (nb) lohtesveisa (se) kälsvets (sv) pienahitsi (fi)

filling knife

See "spattle, spatula, filling knife".

fillings, filler material

fyllmiddel (nb) notkkaldat, deavddádas (se) fyllmedel (sv) täyteaine (fi)

filter

See "percolate, filter, strain".

filtering, filtration

filtrering (nb) duobussillen (se) filtrering (sv) suodatus (fi)

filter, strainer

filter (nb) filttar, duobussilli (se) filter (sv) suodatin (fi)

filtration

See "filtering, filtration".

fin

See "rib, fin, web, gill".

fine-grained

finkorna (nb) fiidnagortnat (se) finkornig (sv) hienojyväinen, -rakeinen (fi)

fine thread

fingjenge (nb) fiidnajeaŋga (se) fingänga (sv) hienokierre, toajakierre (fi)

finish

finbearbeiding (nb) fiidnagieđahallan, loahppagieđahallan (se) färdigbehandling, slutbehandling (sv) hienotyöstö, viimeistely (fi)

finish

See "polish, finish, dress, clean".

finishing, machining

bearbeiding[nb], tilarbeiding[nn] (nb) gieđahallan, ávdnen (se) bearbetning (sv) työstö (fi)

finishing stick

See "whetstone, finishing stick".

finishing tap

See "plug tap, finishing tap, bottoming tap".

fire extinguisher

brannslokkingsapparat (nb) jáddadanapparáhta (se) brandsläckare (sv) sammutin (fi)

fire hose

brannslange (nb) jáddadanšláŋŋa (se) brandslang (sv) paloletku (fi)

fish joint, junction plate, but strap

laskskjøt[nb], laskskøyt[nn] (nb) stielkalakta (se) skarvförbindning (sv) limittäisliitos, limiliitos, palstasovite (fi)

fit, fitment

pasning (nb) heivehus (se) passning (sv) sovite (fi)

fitment

See "fit, fitment".

fittings, armature, fixture

armatur (nb) armatuvra (se) armatur (sv) armatuuri, varuste (fi)

fittings, mountings, armature, garniture

beslag, panel (nb) skoađas (se) beslag (sv) hela, heloitus, raudoitus (fi)

fit, tolerance of fit

pressmonn, pressmonn (nb) deaddemunni (se) pressmån (sv)

fix

See "fasten, tighten, fix".

fixing screw

See "set screw, fixing screw".

fixture

See "fittings, armature, fixture".

flange

flens (nb) gearsi, jargŋa (se) fläns (sv) laippa (fi)

flange

See "fold, flange, seam, joint".

flanging, edge raising

utkraging, kraging (nb) ravdadeapmi, čeabetgalljideapmi (se) kragning (sv) venytys, laipoitus (fi)

flank

See "clearance surface, flank".

flank of thread

See "side of thread, flank of thread".

flap, damper, valve disc

spjell (nb) muddenbelle (se) spjäll (sv) säätöpelti (fi)

flare nut spanner

See "flare nut wrench, flare nut spanner".

flare nut wrench, flare nut spanner

åpen ringnøkkel[nb], open ringnøkkel[nn] (nb) rabas nástečoavdda, rabas gierdočoavdda (se) öppen ringnyckel (sv) avolenkkiavain (fi)

flash light

blinklys, varsellys (nb) šleađggočuovga (se) blinkers (sv) vilkkuvalo (fi)

flat

plan, flat (nb) duolbbas (se) plan (sv) taso, tasainen (fi)

flat chisel

flatmeisel (nb) duolbaluovččan (se) flatmejsel (sv) lattataltta (fi)

flat file, blunt file, safe-edge file

ansatsfil (nb) steallefiilu (se) ansatsfil (sv) lattaviila (fi)

flat nose pliers

See "flat pliers, flat nose pliers".

flat pliers, flat nose pliers

nebbtang[nb], spisstang[nb], flattang[nb], nebbtong[nn], spisstong[nn], flattong[nn] (nb) duolbabasttat, njunnebasttat (se) flacktång (sv) lattapihdit (fi)

flat steel

flatstål, flatjern (nb) duolbastálli (se) plattstål (sv) lattateräs (fi)

flat thread, square thread

firkantgjenge (nb) njealječiegatjeaŋga, duolbajeaŋga (se) flatgänga (sv) lattakierre (fi)

flat transmission belt

flatreim (nb) duolbaruopma (se) flatrem (sv) lattahihna (fi)

flexible

See "elastic, flexible".

float ball

See "float, float ball".

float, float ball

flottør (nb) govddon (se) flottör (sv) uimuri (fi)

flotation

flotasjon (nb) govdudeapmi (se) flotation (sv) vaahdotus (fi)

flow direction

See "direction of current, flow direction".

flowmeter

gjennomstrømmingsmåler[nb], gjennomstrøymingsmålar[nn]

(nb) čađagolganmihtádas (se) flödesmätare (sv) virtausmittari (fi)

flue, outlet, vent pipe, hood

avsug, avtrekk (nb) njaman, geson (se) avsugning (sv) poistoimuri (fi)

fluidizer

See "flux, soldering paste, fluidizer".

flute

See "knurl, flute, groove, ripple".

fluted pin, knurled pin

riflepinne, sporstift (nb) rihkkosággi, luoddanibba (se) uratappi, pyälletty tappi (fi)

flux

fluks (nb) fluksa (se) flöde (sv) magneettivuo (fi)

flux density

flukstetthet[nb], flukstettleik[nn] (nb) fluksasuhkodat (se) flödestäthet (sv) magneettivuon tiheys (fi)

flux, soldering paste, fluidizer

flussmiddel (nb) šolgenávnnas, šolggas (se) flussmedel (sv) juoksute, juotostahna (fi)

flywheel

svinghjul (nb) dássenjuvla (se) svänghjul (sv) vauhtipyörä (fi)

fold

knekke, bukke (nb) máhccut (se) vecka (sv) taivuttaa, taittaa (fi)

fold, flange, seam, joint

fals (nb) bonte, sponta (se) fals (sv) huullos, uurre (fi)

folding machine, trimming machine

knekkemaskin, bukkemaskin (nb) máhccunmášiidna (se) falsmaskin (sv) särmäyskone, taivutuskone, laskostuskone (fi)

folding metre rule

See "inch rule, folding metre rule".

fold, rebate, rabbet

false (nb) bontet, spontet (se) falsa (sv) taivuttaa (fi)

forage harvester

forhøster[nb], forhaustar[nn] (nb) fuođđarrájan (se) fälthack (sv) niittosilppuri (fi)

forceps

See "pinzers, tweezers, forceps".

forecarriage

forstilling (nb) ovdastelle (se) framvagn (sv) etuvaunu, ohjausteline (fi)

forge

smi (nb) dáhkut (se) smida (sv) takoa (fi)

forge

See "smithy, forge".

forgeable, malleable, ductile

smibar (nb) dáhkohahtti (se) smidbar (sv) taottava (fi)

forge hearth

smiesse (nb) álvi (se) smideshärd, ässja (sv) ahjo (fi)

forger

See "smith, blacksmith, forger".

forge tongs

smitang[nb], smitong[nn] (nb) bádjebasttat (se) smidestång (sv) pajapihdit (fi)

forge-welding

essesveising, smisveising (nb) bádjesveisen (se) smidessvetsning (sv) pajahitsaus (fi)

forging hammer

See "sledge hammer, forging hammer".

fork lift truck

gaffeltruck (nb) gaffaltrukka (se) gaffeltruck (sv) haarukkatrukki (fi)

forming

plastisk bearbeiding (nb) plástalaš gieđahallan (se) plastisk bearbetning (sv) muovaava työstö, muovaus (fi)

found

See "cast, mould, found".

foundation, base

fundament, sokkel (nb) vuloštus, stealládat (se) fundament (sv) perustus, laiteperustus (fi)

foundry

støperi[nb], støyperi (nb) leikehat (se) gjuteri (sv) valimo (fi)

four-stroke engine

firetaktsmotor (nb) njealjedávttatmohtor (se) fyrtaktsmotor (sv) nelitahtimoottori (fi)

four wheel bike

firehjulssykkel, firehjuling (nb) njealjejuvllat (sihkkal) (se) fyrhjuling (sv) mönkijä (fi)

frame

See "rib, frame".

free cutting steel, free machining steel

automatstål (nb) automáhtastálli (se) automatstål (sv) automaattiteräs (fi)

free machining steel

See "free cutting steel, free machining steel".

frequency

frekvens (nb) dávji (se) frekvens (sv) taajuus (fi)

friction

friksjon (nb) goahca, itna (se) friktion (sv) kitka (fi)

friction coupling

friksjonskopling (nb) goahcalakta (se) friktionskoppling (sv) kitkakytkin (fi)

friction tape

See "insulating tape, friction tape".

front axle

foraksel, framaksel (nb) ovdaáksil (se) framaxel (sv) etuakseli (fi)

front wheel drive

framhjulsdrevet[nb], -dreven[nn], forhjuls- (nb) ovdajuvlageassin (se) framhjulsdriven (sv) etuvetoinen (fi)

fuel

brennstoff, drivstoff, brensel (nb) boaldámuš (se) bränsle (sv) polttoaine (fi)

funnel

trakt, trekt (nb) reakta, ráhtte, skuibi (se) tratt (sv) suppilo (fi)

fuse

sikring (nb) sihkkarasti (se) säkring (sv) sulake, varoke (fi)

galvanization

See "zink-plating, galvanization".

garbage can, trash barrel, dustbin

søppeldunk (nb) doapparlihtti, ruskalihtti (se) avfallskärl (sv) roskalaatikko, roskasäiliö (fi)

garniture

See "fittings, mountings, armature, garniture".

gas igniter

gnisttenner[nb], gneistetennar[nn], gass- (nb) čuonancahkkehat (se) gaständare (sv) kaasunsytytin (fi)

gasket, jointing, packing

pakning (nb) spintu, bagadas, bagaldat (se) packning (sv) tiiviste (fi)

gasoline motor

See "petrol engine, gasoline motor".

gasoline, petrol, benzine

bensin (nb) bensiidna (se) bensin (sv) bensiini (fi)

gate valve, slide valve

sluseventil (nb) buođđoventiila (se) slussventil, slidventil (sv) luistiventtiili (fi)

gauge

tolk (nb) dulka (se) tolk (sv) tulkki (fi)

gauge

See "measuring tool, gauge".

gauge block

passbit (nb) dárkkistanbihttá (se) passbit (sv) mittapala (fi)

gauge block

parallellkloss (nb) paralleallabihttá (se) parallellbit (sv) suuntaispala (fi)

gauge, jig, template

mal, sjablon, lære (nb) málle (se) schablon, mall (sv) malline, tulkki (fi)

gear

gir (nb) molsson, giira (se) växel (sv) vaihde (fi)

gearbox, transmission

girkasse (nb) molssonkássa, giirakássa (se) växellåda (sv) vaihteisto (fi)

gearing, transmission, exchange

utveksling (nb) molsa (se) utväxling (sv) välitys (fi)

gear wheel, cogwheel, toothed wheel

tannhjul (nb) bátnejuvla, bátneráhtis (se) kugghjul (sv) hammaspyörä (fi)

generator, dynamo

generator, dynamo (nb) generáhtor (se) generator, dynamo (sv) generaattori, laturi, dynamo (fi)

gill

See "rib, fin, web, gill".

glow

See "anneal, glow".

glow

glø, gløde (nb) áhcagastit (se) glöda (sv) hehkua (fi)

glue

lime (nb) liibmet (se) limma (sv) liimata (fi)

glue, cement

lim (nb) liibma (se) lim (sv) liima (fi)

governing, regulation, adjustment

regulering (nb) reguleren (se) reglering (sv) säätö, asetus (fi)

governor

See "regulator, governor".

grab tongs

griptang[nb], låsetang[nb], gripetong[nn], låsetong[nn] (nb) čárvendoaŋggat, čárvenbasttat (se) griptång (sv) lukkopihdit (fi)

graduated cylinder

måletrommel (nb) mihtidanfárppal (se) mättrumma (sv) mittarumpu (fi)

grain size

kornstørrelse[nb], kornstorleik[nn] (nb) gordnesturrodat (se) kornstorlek (sv) raekoko (fi)

grease

See "lubricate, grease".

grease gun

fettpresse[nb], feitt-presse[nn] (nb) vuoiddasbožán (se) fettspruta (sv) voitelupuristin, voiteluruisku (fi)

grease nipple

smørenippel[nb], smørjenippel[nn] (nb) vuoidannihppel (se) smörjnippel (sv) voitelunippa (fi)

grinding disc

See "grinding wheel, grinding disc".

grinding gauge

slipelære (nb) šliipenmálle (se) teräkulma tulkki (fi)

grinding machine

slipemaskin (nb) šliipenmašiidna (se) slipmaskin (sv) hiomakone (fi)

grinding pin

slipestift (nb) šliipennibba (se) slipstift (sv) hiomakara (fi)

grinding, sharpening

sliping (nb) šliipen (se) slipning (sv) hiominen (fi)

grinding wheel

See "grind stone, grinding wheel".

grinding wheel, grinding disc

slipeskive (nb) šliipenskearru (se) slipskiva (sv) hiomalaikka (fi)

grind, sharpen

slipe (nb) šliipet (se) slipa (sv) hioa (fi)

grind stone, grinding wheel

slipestein (nb) šliipa (se) slipsten (sv) hiomakivi, tahko (fi)

grip

See "haft, shaft, handle, grip".

groove

fuge (nb) gurradit (se) foga (sv) railon valmistus (fi)

groove

fuge (nb) sálvu, sveisengurra (se) fog (sv) railo, välys (fi)

groove

See "knurl, flute, groove, ripple".

ground

See "earth, ground, bond".

grounding

See "earthing, grounding, bonding".

gudgeon, pivot, journal, trunnion

tapp (nb) culci, dáhppa (se) tapp (sv) tappi (fi)

guide bar

sverd (nb) miehkki (se) svärd (sv) laippa, terälaippa (fi)

hack saw, bow saw

baufil, skjærfil[nb], skjerfil[nn], buesag[nb], bogesag[nn] (nb) dávgesahá, ruovdesahá (se) bågfil, bågsåg (sv) rautasaha, kaarisaha (fi)

haft, shaft, handle, grip

skaft, handtak, håndtak[nb] (nb) nađđa, geavja (se) skaft, handtag (sv) kahva, varsi, kädensija (fi)

haft, stock

skjefte (nb) nađđadit (se) skafta (sv) varttaa (fi)

hair pin spring

fjærsplint[nb], fjørsplint[nn], hårnål (nb) váfistansoađis, vuoktasoađis (se) sakka (fi)

half finished goods, semi-products

halvfabrikat (nb) bealledagahat (se) halvfabrikat (sv) puolivalmiste (fi)

half-round file

halvrundfil (nb) beallejorbafiilu (se) halvrundfil (sv) puolipyöreä viila (fi)

hammer

hammer[nb], hammar[nn] (nb) veažir (se) hammare (sv) vasara (fi)

hammer drill, percussion drilling machine

slagboremaskin (nb) časkinbovrenmašiidna (se) slagborrmaskin (sv) iskuporakone (fi)

handle

See "crank, handle".

handle, lever, bar

hendel, spak (nb) geavja (se) spak (sv) vipu (fi)

hand pallet truck

jekketralle (nb) loktenvávdna (se) gaffellyftvagn (sv) nostovaunu (fi)

hand riveter

See "rivet tool, hand riveter".

hand vice

See "filing vice, hand vice".

hand wheel, steering wheel

ratt (nb) ráhtta (se) ratt (sv) ohjauspyörä (fi)

hanger

See "rod, hanger, strutting".

hard disc

platelager, harddisk (nb) garraskearru, garramuitu (se) hårddisk (sv) kiintolevy, kovalevy (fi)

hardening furnace, tempering furnace

herdeovn[nb], herdeomn[nn] (nb) buoššodanomman (se) härdugn (sv) karkaisu-uuni (fi)

hardening, tempering

herding (nb) buoššodeapmi, gállen (se) härdning (sv) karkaisu (fi)

harden, temper

herde (nb) buoššodit, gállet (se) härda (sv) karkaista (fi)

hard metal

hardmetall (nb) garrametálla (se) hårdmetall (sv) kovametalli (fi)

hardness

hardhet[nb], hardleik[nn] (nb) garasvuohta, garradat (se) hårdhet (sv) kovuus (fi)

harrow

harv (nb) hárva (se) harv (sv) äes, karhi (fi)

harvester

See "mower, reaper, harvester, mowing machine".

hay press

høyballepresse (nb) suoidnespábbačárvvon (se) rundbalspress, balpress (sv) paalain (fi)

hay rake, buckrake

høysvans (nb) traktordáigu (se) hösvans (sv) heinähäntä (fi)

head stock, mandrel stock, spindle

spindeldokke (nb) jorredoalan (se) spindeldocka (sv) karapylkkä (fi)

hearing protection

See "ear protection, hearing protection".

heat

varme opp, oppvarme (nb) báhkadit (se) värma, upphetta, uppvärma (sv) kumentaa (fi)

heating value, calorific value

brennverdi, varmeverdi (nb) liekkasárvu (se) värmevärde (sv) paloarvo, lämpöarvo (fi)

heatproof, heat resistant

varmefast, varmebestandig[nb] (nb) báhkkagierdevaš, báhkasgierdil (se) värmebeständig (sv) lämmönkestävä (fi)

heat resistant

See "heatproof, heat resistant".

heat treatment

varmebehandling (nb) báhkkagieđahallan (se) värmebehandling (sv) lämpökäsittely (fi)

heavy force fit, heavy shrink fit

presspasning (nb) deaddeheivehus (se) presspassning (sv) puristustiukkuus, luja pakotustiukkuus (fi)

heavy shrink fit

See "heavy force fit, heavy shrink fit".

height of centres

senterhøyde[nb], senterhøgd[nn] (nb) guovddášallodat (se) dubbhöjd (sv) kärkikorkeus (fi)

hexagonal headed bolt

sekskantskrue (nb) guđačiegat skruvva, guđaborat skruvva (se) sexkantskruv (sv) kuusioruuvi (fi)

hexagon nipple

ansatsnippel (nb) steallenihppel (se) sexkantnippel (sv) kaksoisnippa (fi)

hexagon wrench, allen key

sekskantnøkkel, utvendig nøkkel, unbrakonøkkel (nb) guđačiegat čoavdda, guđaborat čoavdda (se) sexkantnyckel (sv) kuusioavain (fi)

high pressure

høytrykk[nb], høgtrykk[nn] (nb) alladeatta (se) högtryck (sv) korkeapaine (fi)

high pressure cleaner

høytrykkspyler[nb], høgtrykkspylar[nn] (nb) alladeattacirggon (se) högtrycksaggregat (sv) painepesuri, korkeapainepesulaite (fi)

high-speed steel

hurtigstål, snøggstål[nn] (nb) spáitastálli (se) snabbstål (sv) pikateräs (fi)

hinge

hengsel, hengsle (nb) gižaldat (se) gångjärn (sv) sarana (fi)

hoist pulley

See “tackle, hoist pulley”.

holder

holder[nb], haldar[nn] (nb) doalan (se) hållare (sv) pidin, kiinnitin (fi)

hollow punch

hullpipe[nb], holpipe[nn] (nb) ráiganbiipu (se) hålstans, huggpipa (sv) reikämeisti, lävistin (fi)

hone

hone (nb) hunet (se) hona (sv) hoonata, laahia (fi)

hone

bryne, heine (nb) sadjit, dávžat (se) bryna (sv) hioa, teroittaa (fi)

hood

See “flue, outlet, vent pipe, hood”.

hook spanner

hakenøkkel (nb) roahkkečoavdda (se) haknyckel (sv) hakaavain (fi)

horisontal, level

vannrett[nb], vassrett[nn], horisontal (nb) láskut, veallut (se) vågrät, horisontell (sv) vaakasuora (fi)

horisontal milling machine, plain milling machine

horisontalfresemaskin (nb) vealujursanmašiidna (se) horisontalfräsmaskin, planfräsmaskin (sv) tasojyrsinkone (fi)

horn, signal horn

horn, signalhorn (nb) jietnadorve (se) signalhorn (sv) torvi, äänitorvi (fi)

hose

slange (nb) šláŋŋa (se) slang (sv) letku (fi)

hose clamp

See "hose clip, hose clamp".

hose clip, hose clamp

slangeklemme (nb) šláŋŋačárvvon (se) slangklämma (sv) letkunkiristin (fi)

hub, nave, boss

nav (nb) juvlaguovddáš, juvlanáhpi (se) nav (sv) napa (fi)

hydraulic

hydraulisk (nb) hydraulalaš (se) hydraulisk (sv) hydraulinen (fi)

hydraulics

hydraulikk (nb) hydraulihkka (se) hydraulik (sv) hydrauliikka, neste- (fi)

idle running

See "idling, no-load, idle running".

idling, no-load, idle running

tomgang (nb) guorusnatjohtin (se) tomgång (sv) joutokäynti (fi)

ignition coil, spark coil

tennspole, coil (nb) cahkkehangiesttus (se) tändspole (sv) sytytyspuola (fi)

ignition plug

See "spark plug, ignition plug".

ignition system

tenningssystem, tenningsanlegg, tenning (nb) cahkkehanrusttet (se) tändsystem (sv) sytytyslaitteisto (fi)

illuminance

belysning, illuminans (nb) čuvgehus (se) belysningsstyrka, illuminans (sv) valaisu, valaistusvoimakkuus (fi)

impact resistance, impact strength

slagseighet[nb], slagseigleik[nn] (nb) časkindávgadat (se) slagseghet (sv) iskusitkeys (fi)

impact strength

See "impact resistance, impact strength".

imperial thread

See "inch thread, imperial thread".

impulse

impuls (nb) impulsa (se) impuls (sv) impulssi (fi)

inch rule, folding metre rule

meterstokk, tommestokk (nb) mehtarmihtádas, dumámihtádas (se) mätstock, tumstock (sv) nivelmitta (fi)

inch thread, imperial thread

tommegjenge (nb) dumájeaŋga (se) tumgänga (sv) tuumakierre (fi)

indexable insert

vendeskjær[nb], vendeskjer[nn] (nb) gomuhanávju, dearrebihttá (se) vändskär (sv) vaihdettava kovametalliteräpala (fi)

index plate, dividing plate

deleskive (nb) juohkinskearru (se) delningsskiva (sv) jakolevy (fi)

indicator

viser[nb], visar[nn] (nb) viisár (se) visare (sv) osoitin (fi)

induce

indusere (nb) induseret (se) inducera (sv) indusoida (fi)

induction

induksjon (nb) indukšuvdna (se) induktion (sv) induktio (fi)

inductive

induktiv (nb) induktiivalaš (se) induktiv (sv) induktiivinen (fi)

inert gas

inaktiv gass, inert gass (nb) doaimmahis gássa (se) inert gas (sv) inerttikaasu (fi)

inert gas

See “noble gas, inert gas”.

inflammable

See “combustible, inflammable”.

ingot, bar

barre (nb) bárra (se) tacka (sv) harkko (fi)

injection

injeksjon, innsprøyting (nb) cirgun (se) insprutning (sv) ruiskutus (fi)

injector

injektor (nb) cirgoventiila (se) insprutningsventil (sv) ruiskutusventtiili (fi)

inlet

See “suction, inlet”.

inner race

See “inner ring, inner race”.

inner ring, inner race

innerring (nb) sisriekkis (se) innerring, inre lagerring (sv) laakerin sisäkehä (fi)

inspection, control

kontroll (nb) dárkkistus (se) kontroll (sv) tarkastus (fi)

install

installere (nb) sajáiduhttit, sajálduhttit (se) installera (sv) asentaa, installoida (fi)

insulate

See "isolate, insulate".

insulating tape, friction tape

isolerband, isolasjonstape (nb) errenbáddi (se) isolerband (sv) eristysnauha (fi)

insulation

See "isolation, insulation".

insulator, separator

isolator (nb) erreldat (se) isolator (sv) eriste, eristin (fi)

interface

grensesnitt (nb) lakta (se) gränssnitt (sv) liitäntä (fi)

interference

interferens (nb) interferensa (se) interferens (sv) interferenssi (fi)

intermediate shaft

See "countershaft, intermediate shaft".

intermittent

See "periodical, cyclic, intermittent, pulsating".

intermittent light

blinklys, retningslys (nb) ravkečuovga (se) körriktningsvisare, blinker, blinkljus (sv) vilkkuvalo (fi)

interrupter

See "switch, circuit breaker, interrupter".

interval

intervall (nb) áigodat (se) intervall (sv) väliaika, väli (fi)

iron

jern (nb) ruovdi (se) järn (sv) rauta (fi)

iron bar

See "crow bar, lever, iron bar".

iron ore

jernmalm (nb) ruovdemálbma (se) järnmalm (sv) rautamalmi (fi)

isolate, insulate

isolere (nb) erret (se) isolera (sv) eristää (fi)

isolation, insulation

isolasjon (nb) erren, guđju (se) isolation (sv) eristys (fi)

jack

jekk (nb) duŋke, jeahkka (se) domkraft (sv) nosturi, tunkki (fi)

jalt

See "close a rivet, upset, clinch, jalt".

jaw chuck

bakkskive (nb) oalulgiddehat (se) backskiva (sv) leukaistukka (fi)

jaw, die

bakke (nb) oalul (se) back (sv) kiinnitysleuka (fi)

jet

See “nozzle, spray nozzle, jet”.

jig

See “gauge, jig, template”.

jig

jigg (nb) valljudanmálle (se) jigg (sv) ohjain, malline, jiki (fi)

joiner, carpenter

tømrer[nb], tømrar[nn] (nb) huksejeaddji (se) timmerman (sv) puuseppä, kirvesmies (fi)

joiner's saw

snekkersag[nb], snikkarsag[nn] (nb) fiellosahá, snihkkársahá (se) snickarsåg (sv) käsisaha, nikkarinsaha (fi)

joining

sammenføyning[nb], samanføying[nn] (nb) oktiibidjan, oktiigidden (se) sammanfogning (sv) liitos (fi)

joint

See “fold, flange, seam, joint”.

joint hook

See “square, joint hook”.

jointing

See “gasket, jointing, packing”.

journal

See “gudgeon, pivot, journal, trunnion”.

journal bearing

See “radial bearing, journal bearing”.

junction plate

See “fish joint, junction plate, but strap”.

key bed

See “keyway, key groove, key bed”.

keyboard

tastatur (nb) boallobeavdi (se) tangentbord (sv) näppäimistö (fi)

key groove

See “keyway, key groove, key bed”.

keyway cutter

kilsporfres (nb) lohtegurrajurssan (se) kilspårsfräs (sv) kiilauranjyrsin (fi)

keyway, key groove, key bed

kilspor (nb) lohteluodda (se) kilspår (sv) kiilaura (fi)

kick, recoil, backfire

tilbakeslag, bakslag (nb) galkan (se) bakslag (sv) takatuli (fi)

knee

See “bend, elbow, angle, knee”.

knife

kniv (nb) niibi (se) kniv (sv) veitsi (fi)

knife tool

knivstål (nb) niibestálli (se) knivstål (sv) veitsiterä (fi)

knock rating

See "octane number, knock rating".

knurled pin

See "fluted pin, knurled pin".

knurl, flute, groove, ripple

rille, ripe, rifle (nb) rihkku (se) spår, rilla (sv) pykälä, ura, uritus, ympyräura (fi)

lamp bulb

See "bulb, lamp bulb".

lapping

lepping (nb) fierdin, snađđen (se) läppning (sv) hierto (fi)

lateral edge, cross edge

tverregg (nb) doaresávju (se) tväregg (sv) poikkisärmä (fi)

lathe

dreiebenk (nb) várve, várvenbeaŋka, jorahanbeaŋka, vátnanbeaŋka (se) svarv (sv) sorvi (fi)

lathe bed

See "bed, lathe bed".

lathe carrier disc

medbringerskive[nb], medtakarskive[nn] (nb) jorahanskearru (se) medbringarskiva (sv) vääntiölaikka (fi)

lathe tool, turning tool

dreiestål (nb) várvenstálli (se) svarvstål (sv) sorvinterä (fi)

lead accumulator, lead battery

blyakkumulator, blybatteri (nb) ladjoakkumuláhtor (se) blyackumulator, blybatteri (sv) lyijyakku (fi)

lead battery

See "lead accumulator, lead battery".

lead-free gasoline / petrol

See "leadless, lead-free gasoline / petrol".

leading screw

See "lead screw, leading screw".

leadless, lead-free gasoline / petrol

blyfri bensin (nb) lajuhis bensiidna (se) blyfri bensin (sv) lyijytön bensiini (fi)

lead screw, leading screw

ledeskrue[nb], leieskrue[nn] (nb) jođihanskruvva (se) ledarskruv (sv) johtoruuvi (fi)

leaf spring, plate spring

bladfjær[nb], bladfjør[nn] (nb) duolbadávgi (se) bladfjäder (sv) lehtijousi (fi)

leakage

See "leak, leakage".

leak, leakage

lekkasje, lekk (nb) suođđu (se) läckage (sv) vuoto (fi)

leak, leakage

lekkasje, lekk (nb) vuohču (se) läckage (sv) vuoto (fi)

leftward welding

frasveising[nb], fråsveising[nn], med-, venstre- (nb) eretsveisen (se) motsvetsning (sv) myötähitsaus (fi)

length of stroke, piston travel

slaglengde[nb], slaglengd[nn] (nb) časkinguhkkodat (se) slaglängd (sv) iskunpituus (fi)

level

See "horisontal, level".

level, water level

vater, vaterpass (nb) čáhcečalbmi, váhttar (se) vattenpass (sv) vesivaaka (fi)

lever

See "crow bar, lever, iron bar".

lever

hevarm, hevstang[nb], hevstong[nn], vippe (nb) loktenstággu (se) hävstång (sv) vipu, vipuvarsi (fi)

lever

vektstang[nb], vektstong[nn] (nb) veaktastággu (se) hävstång (sv) vipu (fi)

lid

See "cap, lid, cover".

light metal

lettmetall (nb) geahpametálla (se) lättmetall (sv) kevytmetalli (fi)

limit value

grensemål (nb) rádjemihttu (se) gränsmått (sv) rajamitta (fi)

line voltage

See "mains voltage, line voltage".

lining

See "coat, coating, facing, lining, surfacing".

liquid

væske (nb) njalbi (se) vätska (sv) neste (fi)

live centre

roterende senterspiss[nb], roterande senterspiss[nn] (nb) jorri guovddášnjunni (se) roterande dubb (sv) pyörivä keskiökärki (fi)

load

belastning (nb) noađuheapmi, guorbmádeapmi (se) belastning (sv) kuormitus (fi)

load

belaste (nb) noađuhit, guorbmádit (se) belasta (sv) kuormittaa (fi)

loadbearing capacity

See "carrying capacity, loadbearing capacity".

location

See "situation, site, location".

locking device, catch

sperremekanisme, sperre (nb) giddenmekanisma (se) spärranordning, spärr (sv) säppi, lukitsin, salpa (fi)

locking ring

låsering, seegerring (nb) lássenriekkis, lohkkadanriekkis (se) låsring (sv) lukitusrengas (fi)

locking washer

låseskive (nb) lássenskearru (se) låsbricka (sv) lukkoaluslevy (fi)

locknut

See "counter nut, locknut".

loop

See "bow, clamp, stirrup, loop".

loose jaws

spennblikk (nb) čávgenspelle, čávgenbelle (se) suojapakat, leuansuojukset (fi)

loosen

løs(n)e[nb], løyse[nn] (nb) čoavdit (se) lossa (sv) hellittää (fi)

low pressure, negative pressure, depression

lavtrykk[nb], lågtrykk, undertrykk (nb) vuollegisdeatta, vuolledeatta (se) lågtryck, undertryck (sv) alipaine, matalapaine (fi)

lubricant

smøremiddel[nb], smørjemiddel[nn] (nb) vuoiddas (se) smörjmedel (sv) voiteluaine (fi)

lubricate, grease

smøre[nb], smørje[nn] (nb) vuoidat (se) smörja (sv) voidella, rasvata (fi)

lubricating film

smørefilm[nb], smørjefilm[nn] (nb) vuoiddasassi, vuoiddasfilbma (se) smörjfilm (sv) voiteluainekerros (fi)

lubricating grease

smørefett[nb], smørjefeitt[nn] (nb) vuoiddanas (se) smörjfett (sv) voitelurasva (fi)

lubricating oil

smøreolje[nb], smørjeolje[nn] (nb) vuoiddasolju (se) smörjolja (sv) voiteluöljy (fi)

lug

See "shoulder, collar, lug".

machine

maskinere (nb) mašineret (se) bearbeta (sv) koneistaa (fi)

machine key

See "wedge, machine key".

machining

See "finishing, machining".

machining

maskinering (nb) mašineren (se) maskinbearbetning (sv) koneistaminen (fi)

machining allowance

See "working margin, machining allowance".

magnet

magnet (nb) magnehta (se) magnet (sv) magneetti (fi)

magnetic field

magnetfelt, magnetisk felt (nb) magnehtagieddi (se) magnetfält (sv) magneettikenttä (fi)

magnetoelectric ignition

magnettenning (nb) magnehtacahkkeheapmi (se) magnettändning (sv) magneettisytytys (fi)

main cutting edge

See "major cutting edge, main cutting edge".

mains, power network, power grid

lysnett (nb) čuovgafierpmádat (se) elnät (sv) sähköverkko (fi)

mains voltage, line voltage

nettspenning (nb) fierbmegealdda (se) nätspänning (sv) verkkovirta, verkkojännite (fi)

maintenance, service

vedlikehold[nb], vedlikehald[nn], service (nb) fuoladeapmi, fuolahus, bajásdoalahus (se) underhåll, skötsel, service (sv) kunnossapito, huolto (fi)

major cutting edge, main cutting edge

hovedegg[nb], hovudegg[nn] (nb) váldoávju, oaiveávju (se) huvudskäregg, huvudegg (sv) pääleikkuusärmä (fi)

malleable

See "forgeable, malleable, ductile".

malleable cast iron

aduserjern (nb) aduserruovdi (se) aducerjärn (sv) adusoitu valurauta (fi)

mallet

See "sledge hammer, mallet, maul".

mallet, beetle

klubbe (nb) šluppot (se) klubba (sv) nuija (fi)

mandrel stock

See "head stock, mandrel stock, spindle".

manifold

manifold, samlestokk, forgreiningsrør (nb) suorgebohcci (se) manifold, samlingsrör, förgreningsrör (sv) pakosarja, imusarja (fi)

manometer

See "pressure gauge, manometer".

manual

manuell (nb) giehtaanulaš (se) manuell (sv) käsikäyttöinen, käsi- (fi)

manual, operating instructions

bruksanvisning[nb], bruksrettleiing[nn] (nb) geavahančilgehus, atninčilgehus (se) bruksanvisning (sv) käyttöohje (fi)

manufacturing plant

See "factory, manufacturing plant".

manufacturing plant, works, company

bedrift, foretak[nb], føretak[nn] (nb) fitnodat (se) företag (sv) yritys (fi)

margin

See "tolerance, allowance, margin".

marking gauge

rissmål, strekmål (nb) nárbma (se) strykmått, streckmått, ritsmått (sv) suunturi, suuntapiirrin (fi)

marking gauge

See "surface gauge, marking gauge".

material

materiale, emne (nb) ávnnas (se) material, ämne (sv) aine, aines, materiaali (fi)

maul

See "sledge hammer, mallet, maul".

measure

målsette, målsetje[nn] (nb) mihttobidjan (se) måttsätta (sv) mitoittaa (fi)

measure, measuring, measurement

måling (nb) mihtideapmi (se) mätning (sv) mittaaminen, mittaus (fi)

measurement

See "measure, measuring, measurement".

measure out

See "dimension, measure out".

measuring

See "measure, measuring, measurement".

measuring surface

målekjeft (nb) mihtidanolggoš, mihtidanoalul (se) mätyta (sv) mittauspinta (fi)

measuring tape, steel tape rule

måleband, bandmål (nb) mihtidanbáddi (se) bandmått (sv) mittanauha (fi)

measuring tool, gauge

måleverktøy (nb) mihtádas, mihtidanneavvu, mihttár (se) mätverktyg (sv) mittaustyökalu, mittain (fi)

mechanical

mekanisk (nb) mekánalaš, mekanihkkalaš (se) mekanisk (sv) mekaaninen (fi)

mechanician

See "mechanic, mechanician".

mechanic, mechanician

mekaniker[nb], mekanikar[nn], reparatør (nb) mekanihkkár, divodeaddji (se) mekaniker (sv) asentaja, korjaaja (fi)

mechanism

mekanisme (nb) mekanisma (se) mekanism (sv) mekanismi, koneisto (fi)

mechanization

mekanisering (nb) mekaniseren (se) mekanisering (sv) koneellistaminen (fi)

melt

smelte (nb) suddadit (se) smälta (sv) sulattaa (fi)

melt

smelte (nb) suddat (se) smälta (sv) sulaa (fi)

melting point

See "melting temperature, melting point".

melting temperature, melting point

smeltetemperatur, smeltepunkt (nb) šolgantemperatuvra (se) smälttemperatur, smältpunkt (sv) sulamispiste (fi)

melt, molten mass

smelte (nb) šolgu (se) smälta (sv) sulatuserä, sula (fi)

membrane, diaphragm

membran (nb) cuozza, membrána (se) membran (sv) kalvo (fi)

memory

minne (nb) muitu (se) minne (sv) muisti (fi)

mend

See "repair, restore, mend".

menu

meny (nb) fállu (se) meny (sv) valikko (fi)

mesh

See "engagement, mesh".

metal

metall (nb) metálla (se) metal (sv) metalli (fi)

metallic glint

See "metallic lustre, metallic glint".

metallic lustre, metallic glint

metallglans (nb) metállašealgu (se) metallglans (sv) metallinkiilto, -hohde (fi)

metric

metrisk (nb) mehtarlaš (se) metrisk (sv) metrinen (fi)

micrometer

mikrometer (nb) mikromihtádas (se) mikrometer (sv) mikrometri (fi)

microprocessor

mikroprosessor (nb) mikrodoaimmár (se) mikroprocessor (sv) mikroprosessori, mikrosuoritin (fi)

micro switch

mikrobryter[nb], mikrobrytar[nn] (nb) mikrobotkkán (se) mikroströmbrytare (sv) mikrokytkin (fi)

mill

frese (nb) jurssahit, jursat (se) fräsa (sv) jyrsiä (fi)

milling cutter

fres (nb) jurssan (se) fräs (sv) jyrsin (fi)

milling machine

fresemaskin (nb) jursanmašiidna (se) fräsmaskin (sv) jyrsinkone (fi)

minor cutting edge

See "secondary edge, minor cutting edge".

mitre rule

See "bevel rule, mitre rule".

mixer tap

See "mixing battery, mixer tap, mixing cock".

mixing

See "mixture, mixing".

mixing battery, mixer tap, mixing cock

blandebatteri (nb) seaguhanhátna (se) blandare (sv) sekoitin (fi)

mixing chamber

blandekammer (nb) seaguhangámmár (se) blandningskammare (sv) sekoituskammio (fi)

mixing cock

See "mixing battery, mixer tap, mixing cock".

mixture, mixing

mekanisk blanding (nb) mekánalaš seaguhus (se) mekanisk blandning (sv) mekaaninen seos (fi)

model

modell (nb) málle (se) modell (sv) malli (fi)

modular measure, starting point

basismål (nb) vuođđomihttu (se) basmått (sv) perusmitta (fi)

molecule

molekyl (nb) molekyla (se) molekyl (sv) molekyyli (fi)

molten mass

See "melt, molten mass".

molten pool, weld pool

smeltebad (nb) sveisašolgu (se) svetssmälta (sv) hitsisula (fi)

moment

See "torque, moment".

monitor

See "screen, monitor, display".

moped

moped (nb) mopeda (se) moped (sv) mopo, moottoripolkupyörä (fi)

motion, movement

bevegelse[nb], rørsle[nn] (nb) johtu, lihkadeapmi (se) rörelse (sv) liike (fi)

motive force

See "motive power, thrust, motive force".

motive power, thrust, motive force

drivkraft (nb) jođihanfápmu (se) drivkraft (sv) käyttövoima (fi)

motorbike

See "motor cycle, motorbike".

motor-car

See "car, motor-car".

motor cycle, motorbike

motorsykkel (nb) mohtorsihkkal (se) motorcykel (sv) moottoripyörä (fi)

motor, engine

motor (nb) mohtor (se) motor (sv) moottori (fi)

motor protecting switch

motorvernbryter[nb], motorvernbrytar[nn] (nb) mohtorsuodjebotkkon (se) motorskydd (sv) moottorin suojakytkin (fi)

mould

See "cast, mould, found".

mould

form (nb) foarbma, hoarbma (se) form (sv) muotti (fi)

mount

See "assemble, mount".

mountings

See "fittings, mountings, armature, garniture".

mouse

mus (nb) sáhpán (se) mus (sv) hiiri (fi)

mouth piece, nozzle, die

munnstykke (nb) njálbmebihttá, geahčebihttá (se) munstycke (sv) suutin (fi)

movement

See “motion, movement”.

mower, reaper, harvester, mowing machine

slåmaskin (nb) láddjenmašiidna (se) slåttermaskin (sv) niittokone (fi)

mowing machine

See “mower, reaper, harvester, mowing machine”.

muffler

See “silencer, damper, muffler, exhaust box”.

multipower end cutting nippers

See “bolt cutter, multipower end cutting nippers”.

nail

spiker[nb], spikar[nn] (nb) spiikkár (se) spik (sv) naula (fi)

nail puller

spikertrekker[nb], spikartrekkjar[nn], kjerringkjeft (nb) spiikkárgeasán (se) spikutdragare (sv) sorkkarauta, naulanvedin (fi)

nail puller, nippers, pincers

hovtang[nb], hovtong[nn], tversavbiter[nb], tversavbitar[nn] (nb) gazzadoaŋggat, lihtdoaŋggat (se) hovtång (sv) hohtimet (fi)

nave

See “hub, nave, boss”.

needle bearing

nålelager (nb) nálloláger (se) nållager (sv) neulalaakeri (fi)

negative pressure

See “low pressure, negative pressure, depression”.

nippers

See “nail puller, nippers, pincers”.

nipple

nippel (nb) nihppel (se) nippel (sv) nippa, nippeli (fi)

noble gas, inert gas

edelgass (nb) jállogássa, árvogássa (se) ädelgas (sv) jalokaasu (fi)

nodular cast iron

kulegrafittjern, seigjern (nb) jorbagrafihttaruovdi (se) segjärn (sv) pallografiittirauta (fi)

noise

støy (nb) šlápma, válla (se) buller (sv) melu (fi)

no-load

See “idling, no-load, idle running”.

no load voltage, open circuit voltage

tomgangsspenning (nb) guorusnatgealdda (se) tomgångsspänning (sv) tyhjäkäyntijännite (fi)

non-acceptance limit

ikke-gå-grense[nb], ikkje-gå-grense[nn] (nb) garvinrádji (se)

non-return valve

See "check valve, non-return valve".

normalize

normalisere (nb) normaliseret (se) normalisera (sv) normalisoida (fi)

normal position

normalstilling, nøytralstilling (nb) normálasajádat (se) normalläge (sv) normaaliasento (fi)

notch

See "cut, notch, section".

nozzle

See "mouth piece, nozzle, die".

nozzle, spray nozzle, jet

dyse (nb) cirggon (se) dysa (sv) suutin (fi)

number of teeth

tanntall[nb], tanntal[nn] (nb) bátnelohku (se) kuggtal (sv) hammasluku (fi)

nut

mutter (nb) muhtter (se) mutter (sv) mutteri (fi)

octane number, knock rating

oktantall[nb], oktantal[nn] (nb) oktánalohku (se) oktantal (sv) oktaaniluku (fi)

oil

olje (nb) olju (se) olja (sv) öljy (fi)

oil bath

oljebad, oljesump (nb) oljolávgu (se) oljebad (sv) öljykylpy, öljyallas (fi)

oil can

oljekanne, smøreoljekanne, smørjeoljekannen (nb) oljogádnu (se) oljekanna (sv) öljykannu, voitelukannu (fi)

oil groove

smørespor[nb], smørjespor[nn] (nb) vuoidangurra, vuoidanluodda (se) smörjspår (sv) voitelu-ura (fi)

oil mist lubricator

(olje)tåkesmører[nb], -smørjar[nn] (nb) mierkavuoidi (se) oljedimbildare (sv) voidesumutin (fi)

open circuit voltage

See "no load voltage, open circuit voltage".

open-end spanner, open-end wrench

fastnøkkel (nb) rabas čoavdda (se) U-nyckel (sv) kiintoavain (fi)

open-end wrench

See "open-end spanner, open-end wrench".

operating instructions

See "manual, operating instructions".

operation

See "steering, control, operation".

operation, running

drift (nb) doaibma (se) drift (sv) käyttö (fi)

ore

malm (nb) málbma (se) malm (sv) malmi (fi)

ore concentrate

See "dressed ore, ore concentrate".

O-ring seal

O-ring (nb) O-riekkis (se) O-ring (sv) O-rengas (fi)

oscillate

See "weave, oscillate".

oscilloscope

oscilloskop (nb) oscilloskohppa (se) oscilloskop (sv) oskilloskooppi (fi)

outboard engine

påhengsmotor, utenbordsmotor[nb], uta(n)bords-[nn] (nb) maŋŋegeašmohtor, fanasmohtor (se) utombordsmotor (sv) perämoottori (fi)

outer race

See "outer ring, outer race".

outer ring, outer race

ytterring (nb) olgoriekkis (se) ytterring (sv) laakerin ulkokehä (fi)

outlet

See "flue, outlet, vent pipe, hood".

outlet, drain

avløp (nb) luoitahat, njielahat (se) avlopp, utlopp (sv) viemäri (fi)

outlet tube

See "drain pipe, outlet tube, sewer pipe".

output unit

ut-enhet[nb], ut-eining[nn] (nb) olggosoassi (se) utenhet (sv)

outside calipers

See "pair of compasses, outside calipers".

overhead travelling crane

See "traverse crane, overhead travelling crane".

over-lapping

overlappskjøt[nb], overlappskøyt[nn] (nb) gokčanlakta, latnalaslakta (se) överlappning (sv) limitys (fi)

overload

overbelastning (nb) liigenoađuheapmi, -ražaheapmi, ilá - (se) överbelastning (sv) ylikuormitus (fi)

overpressure, positive pressure

overtrykk (nb) badjeldeatta (se) övertryck (sv) ylipaine (fi)

oxide scale

See "scale, oxide scale".

packing

See "gasket, jointing, packing".

pair of compasses, outside calipers

krumpasser[nb], fotpasser[nb], krumpassar[nn], fotpassar[nn] (nb) gierdomihtádas (se) krumpassare, krumcirkel (sv) länkiharppi (fi)

parallel connection

See "shunt, shunt switch, parallel connection".

parallel interface

parallelt grensesnitt (nb) buohtalaslakta (se) parallell gränssnitt (sv) rinnakkaisliitäntä (fi)

particle

partikkel (nb) partihkal (se) partikel (sv) hiukkanen (fi)

parts list

stykkliste (nb) oasselistu (se) stycklista, detaljlista (sv) osaluettelo (fi)

password

passord (nb) beassansátni (se) lösenord (sv) tunnussana (fi)

p. control valve

See "pressure reducing valve, p. control valve".

pedal

pedal (nb) duolmman (se) pedal (sv) poljin, jalkapoljin (fi)

peg, pin, plug, stud

plugg (nb) bunci, náhpul (se) plugg (sv) tulppa, sulkutulppa (fi)

pendulum

pendel (nb) dearba, dearbastihkka, heailu, šlimpa (se) pendel (sv) heiluri (fi)

percolate, filter, strain

filtrere (nb) duobussillet (se) filtra (sv) suodattaa (fi)

percussion drilling machine

See "hammer drill, percussion drilling machine".

perforated plate

perforert plate (nb) ráiggaspláhtta (se) perforerad platta (sv) rei'itetty levy, reikälevy (fi)

period

See "cycle, period".

periodical, cyclic, intermittent, pulsating

pulserende[nb], pulserande[nn], periodisk (nb) ravki (se) periodisk (sv) jaksoittainen (fi)

permanent

permanent (nb) bistevaš (se) permanent (sv) kesto- (fi)

perpendicular, vertical

loddrett, vertikal (nb) ceaggut (se) lodrät, vertikal (sv) pystysuora, kohtisuora (fi)

petrol

See "gasoline, petrol, benzine".

petrol engine, gasoline motor

bensinmotor (nb) bensinmohtor (se) bensinmotor (sv) bensiinimoottori (fi)

phase

fase (nb) muddu (se) fas (sv) vaihe (fi)

phase shift

faseforskyving, faseforskuving[nn] (nb) muddosirdu (se) fasförskjutning (sv) vaihesiirto (fi)

phase wear

fasslitasje (nb) ravdaloaktu (se) fasförslitning (sv)

pig iron

råjern (nb) álgguviđá ruovdi (se) tackjärn (sv) harkkorauta (fi)

pillar drilling machine

See "upright drilling machine, vertical drilling machine, pillar drilling machine".

pilot tap

See "taper tap, pilot tap".

pin

See "peg, pin, plug, stud".

pincers

See "nail puller, nippers, pincers".

pin, dowel

pinne (nb) sággi (se) pinne (sv) tappi, vaarna, sokka (fi)

pin hammer

See "straight pane hammer, pin hammer".

pinion

pinjong (nb) pinjoŋga (se) pinjong (sv) pieni vetopyörä (fi)

pinzers, tweezers, forceps

pinsett (nb) pinseahtat (se) pincett (sv) atulat, pinsetit (fi)

pipe

rør, røyr[nn] (nb) bohcci, revre (se) rör (sv) putki (fi)

pipe cutter

rørkutter[nb], rørkuttar[nn] (nb) bohccečuohpan (se) rörkapare (sv) putkenkatkaisin, putkileikkuri (fi)

pipe thread

rørgjenge (nb) bohccejeaŋga (se) rörgänga (sv) putkikierre (fi)

pipe tongs, assembly tong, pipe wrench

rørtang[nb], rørtong[nn] (nb) bohccebasttat, bumpedoaŋggat (se) rörtång (sv) putkipihdit (fi)

pipe wrench

See "pipe tongs, assembly tong, pipe wrench".

piston

stempel (nb) meandi, steamppal (se) kolv (sv) mäntä (fi)

piston ring

stempelring (nb) meanderiekkis (se) kolvring (sv) männänrengas (fi)

piston rod, connecting rod

stempelstang, veivstang, -stong[nn], råde (nb) meandestággu, veivestággu (se) kolvstång, vevstake (sv) männänvarsi, kiertokanki (fi)

piston stroke

See "stroke, piston stroke".

piston travel

See "length of stroke, piston travel".

pitch circle

delesirkel (nb) jorrangierdu (se) rullningscirkel (sv) vierintäympyrä (fi)

pitch circle of bolt holes

hullsirkel[nb], holsirkel[nn] (nb) ráigegierdu (se) hålcirkel (sv)

pitch of thread

See "screw pitch, pitch of thread".

pitch rack

See "rack, pitch rack, toothed bar".

pitting wear

gropslitasje (nb) gohpeloaktu (se) gropförslitning (sv) terän kuoppakuluminen (fi)

pivot

See "gudgeon, pivot, journal, trunnion".

plain bearing

See "sliding bearing, plain bearing".

plain milling cutter

See "cylindrical cutter, plain milling cutter".

plain milling machine

See "horisontal milling machine, plain milling machine".

plane

høvel (nb) heavval (se) hyvel (sv) höylä (fi)

plane

høvle (nb) heavvalasttit (se) hyvla (sv) höylätä (fi)

plane, face

plane (nb) dulbet (se) plana (sv) tasoitta (fi)

planer, shaper, planing machine, shaping machine

høvelmaskin (nb) heavvalmašiidna (se) hyvelmaskin (sv) höyläkone (fi)

planing bench

See "carpenter's bench, planing bench".

planing machine

See "planer, shaper, planing machine, shaping machine".

planing tool, dresser

avretter[nb], avrettar[nn] (nb) njulggonas (se) skärpverktyg, avrivare (sv) teroituslaite, oikaisulaite (fi)

plant, construction works

anlegg (nb) ráhkadus (se) anläggning, byggnadsverk (sv) rakennus, työmaa (fi)

plasma cutter

plasmaskjærer[nb], plasmaskjerar[nn] (nb) plasmačuohpan (se) plasmaskärare (sv) plasmaleikkain, plasmaleikkauslaite (fi)

plastic

See "ductile, plastic".

plastic mallet

plasthammer[nb], plasthammar[nn] (nb) plástaveažir, muovveveažir (se) plasthammare (sv) muovivasara (fi)

plastics

plast (nb) plásta, muovvi (se) plast (sv) muovi (fi)

plate

See "disc, plate, washer".

plate

lamell (nb) lamealla (se) lamell (sv) lamelli, levy (fi)

plate shears

See "sheet metal shears, plate shears".

plate, sheet

plate (nb) pláhtta (se) plåt, platta (sv) levy, laatta (fi)

plate spring

See "leaf spring, plate spring".

pliers

See "tongs, pliers".

plotter

plotter[nb], plottar[nn], grafskriver[nb], grafskrivar[nn] (nb) sárggon (se) kurvskrivare, kurvritare (sv) piirturi (fi)

plough

plog (nb) vealta, pluga (se) plog (sv) aura (fi)

plug

See "peg, pin, plug, stud".

plug

støpsel, plugg (nb) el-culci, sággeguoskkahat (se) stickpropp (sv) tulppaliitin, pistotulppa (fi)

plug tap

See "second tap, plug tap".

plug tap, finishing tap, bottoming tap

bunntapp[nb], botntapp[nn], sluttapp (nb) botnedáhppa (se) gradtapp (sv) viimeistelytappi (fi)

plugwire, spark plug cable

pluggledning[nb], pluggleidning[nn], tennplugg- (nb) ginttaljođas, buncejođas (se) tändstiftskabel (sv) tulpanjohto, sytytustulpan johto (fi)

plumber

rørlegger[nb], røyrleggjar[nn] (nb) bohccebargi (se) rörläggare (sv) putkiasentaja (fi)

plunger

spennstift, rørsplint (nb) bohccesággi (se) fjädrande rörpinne (sv) putkisokka (fi)

pneumatic

pneumatisk (nb) pneumáhtalaš (se) pneumatisk (sv) paineilma-, pneumaattinen (fi)

pneumatic hammer

See "air hammer, pneumatic hammer".

pole

pol (nb) náhpi (se) pol (sv) napa (fi)

polish

polere (nb) šelget, geallat (se) polera (sv) kiillottaa (fi)

polish, finish, dress, clean

pusse (nb) liktet (se) putsa (sv) kiillottaa, puhdistaa (fi)

pollution

forurensning[nb], (for)ureining[n (nb) nuoskkideapmi (se) förorening (sv) epäpuhtaus, saaste (fi)

pop rivet

popnagle, blindnagle (nb) popnávli (se) popnit, blindnit (sv) pop-niitti (fi)

pore forming

poredanning (nb) sáigun (se) porbildning (sv) huokosten muodostuminen (fi)

pore, void

pore, blære (nb) skoavhli, ráhkku (se) por (sv) huokonen (fi)

porous

porøs (nb) silli (se) porös (sv) huokoinen (fi)

positive pressure

See "overpressure, positive pressure".

potential energy

potensiell energi, stillingsenergi (nb) veadjoárja (se) potentiell energi (sv) potentiaalienergia (fi)

powder

pulver (nb) bulvarat, jáffut (se) pulver (sv) pulveri, jauhe (fi)

power drive

kraftuttak (nb) fápmováld-dahat (se) kraftuttag (sv) voimanottolaite, voiman ulosotto (fi)

power grid

See "mains, power network, power grid".

power network

See "mains, power network, power grid".

power saw, chain saw

motorsag (nb) mohtorsahá (se) motorsåg (sv) moottorisaha (fi)

power source, source of current

strømkilde[nb], straumkjelde[nn] (nb) rávdnjegáldu (se) strömkälla (sv) virtalähde (fi)

power stroke, working stroke

arbeidsslag, arbeidstakt (nb) bargodákta (se) arbetstakt (sv) työtahti (fi)

power transmission

kraftoverføring (nb) fápmosirdin (se) kraftöverföring (sv) voimansiirto (fi)

precision, accuracy

presisjon, nøyaktighet[nb], grannsemd[nn] (nb) dárkivuohta, aiddolašvuohta (se) precision, noggrannhet (sv) tarkkuus, täsmällisyys (fi)

prepare, process, treat

bearbeide[nb], tilarbeide[nn] (nb) gieđahallat, ávdnet (se) bearbeta (sv) työstää, valmistaa (fi)

press

presse (nb) deaddit (se) pressa (sv) painaa, puristaa (fi)

press

presse (nb) deattán (se) press (sv) puristin, puristuskone (fi)

pressure gauge, manometer

trykkmåler[nb], trykkmålar[nn], manometer (nb) deattamihtádas, manomehtar (se) manometer (sv) painemittari, manometri (fi)

pressure governor

See "pressure regulator, pressure governor".

pressure reducing valve, p. control valve

trykkbegrensingsventil (nb) deattarádjenventiila (se) tryckreduceringsventil (sv) paineenalennusventtiili (fi)

pressure regulator, pressure governor

trykkregulator (nb) deattareguláhtor (se) tryckregulator (sv) paineensäädin (fi)

pre-stressing, pre-tensioning

forspenning (nb) ovdagealdda, álgogealdda (se) förspänning (sv) esijännitys (fi)

pre-tensioning

See "pre-stressing, pre-tensioning".

primary

primær (nb) vuođđo-, primára, álgo- (se) primär (sv) ensisijainen, pää- (fi)

primary circuit

primærkrets[nb], primærkrins[nn] (nb) vuođđobiire, álgobiire (se) primärkrets (sv) pääpiiri, ensiöpiiri (fi)

printer

skriver[nb], skrivar[nn] (nb) čálán (se) skrivare (sv) kirjoitin, tulostin (fi)

probe

See "detector, sensor, probe".

process

See "prepare, process, treat".

process

prosess (nb) mannolat, proseassa (se) process (sv) prosessi, tapahtumasarja (fi)

processor

prosessor, regneenhet[nb], rekneeining[nn] (nb) doaimmár (se) processor (sv) prosessori, suoritin (fi)

program

programmere (nb) prográmmeret (se) programera (sv) ohjelmoida (fi)

programme

program (nb) prográmma (se) program (sv) ohjelma (fi)

propeller

propell (nb) propealla (se) propeller (sv) potkuri (fi)

proportion of ingredients, ratio of mixture

blandingsforhold (nb) seaguhusgorri (se) blandningsförhållande (sv) seossuhde (fi)

protective equipment

verneutstyr (nb) sudjenneavvut (se) skyddsutrustning (sv) suoja-asu, suojavarusteet (fi)

protective helmet

vernehjelm (nb) sudjenjealbma (se) skyddshjälm (sv) suojakypärä (fi)

protective spectacles

vernebriller (nb) sudjenláset (se) skyddsglasögon (sv) suojalasit (fi)

puller, wheel puller

avtrekker[nb], avtrekkjar[nn], ters, avdrager[nb], avdragar[nn] (nb) geasán (se) avdragare (sv) ulosvedin, irrotin (fi)

pulley

See “rowel, pulley, caster”.

pulley, belt pulley

reimskive (nb) ruopmaskearru (se) remskiva (sv) hihnapyörä (fi)

pulley block

See “block, pulley block”.

pulsating

See “periodical, cyclic, intermittent, pulsating”.

pulse

puls (nb) ravkkas (se) puls (sv) pulssi, syke (fi)

pump

pumpe (nb) bumpa (se) pump (sv) pumppu (fi)

pump

pumpe (nb) bumpet (se) pumpa (sv) pumpata (fi)

punch

See “centre punch, punch”.

punching machine, stamping machine

stansemaskin (nb) vadjanmašiidna (se) stansmaskin (sv) meistauskone, lävistyskone (fi)

punching tool, stamp(er)

stanseverktøy (nb) vajan (se) stansverktyg (sv) meisti (fi)

punch, revolving punch pliers

hulltang[nb], holtong[nn] (nb) ráiganbasttat (se) håltång (sv) reikäpihdit (fi)

punch, stamp

stanse, lokke (nb) vadjat (se) stansa (sv) meistää, lävistää (fi)

push button

trykknapp (nb) deaddinboallu (se) tryckknapp (sv) painike (fi)

putty, fill

sparkle (nb) sparkelastit (se) spackla (sv) siloittaa (fi)

quenching, chilling

bråkjøling (nb) roahtačoaskudeapmi (se) störtkylning (sv) karkaisusammutus (fi)

rabbet

See "fold, rebate, rabbet".

rabbet, scarf

falsskjøt[nb], falsskøyt[nn] (nb) bontelakta, spontalakta (se) fals (sv)

race

ruse (nb) rassehit, guorusrassehit (se) rusa (sv) ryntäyttää (fi)

rack

See "stand, rack, frame".

rack, pitch rack, toothed bar

tannstang, tannstong[nn] (nb) bátnestággu (se) kuggstång (sv) hammastanko (fi)

radial

radiell, radial (nb) radiála, suonjarháltásaš (se) radiell, radial (sv) radiaalinen, säteis-, säteittäinen (fi)

radial bearing, journal bearing

radiallager (nb) radialláger, suonjarláger (se) radiallager (sv) säteislaakeri (fi)

radial drilling machine

radialboremaskin (nb) radialbovrenmašiidna, suonjarbovrenmašiidna (se) radialborrmaskin (sv) säteisporakone (fi)

radiator

radiator (nb) radiáhtor (se) kylare (sv) jäähdytin (fi)

rake, tool cutting edge inclination

hellingsvinkel (nb) smilčečiehka (se) lutningsvinkel (sv) kallistuskulma (fi)

rasp

See "coarse file, rasp".

ratchet handle, ratchet wrench

skralle, smelle (nb) sŋiran, coakcečoavdda (se) spärrnyckel (sv) räikkä, räikkäväännin (fi)

ratchet wheel

palhjul, matehjul (nb) caggejuvla, borahanjuvla (se) steghjul (sv) syöttöpyörä, säppipyörä (fi)

ratchet wrench

See "ratchet handle, ratchet wrench".

rate

See "speed, rate, velocity".

ratio of mixture

See "proportion of ingredients, ratio of mixture".

raw material

råmateriale, råvare (nb) álgoávnnas, luondduviđá ávnnas (se) råmaterial, råvara (sv) raaka-aine (fi)

ream, broach

brotsje, opprømme (nb) galjidit (se) brotscha, upprymma (sv) kalvia, väljentää (fi)

reamer, broach, reaming bit

brotsj, rømmer (nb) galján (se) brotsch, upprymmare (sv) kalvin, väljennin, avennin (fi)

reaming bit

See "reamer, broach, reaming bit".

reaper

See "mower, reaper, harvester, mowing machine".

rear axle

bakaksel (nb) maŋŋeáksil (se) bakaxel (sv) taka-akseli (fi)

rear wheel

bakhjul (nb) maŋŋejuvla (se) bakhjul (sv) takapyörä (fi)

rebate

See "fold, rebate, rabbet".

rechargeable

oppladbar (nb) gealddehahtti (se) laddningsbar (sv) uudelleen ladattava (fi)

recoil

See "kick, recoil, backfire".

record

See "save, record".

rectifier

likeretter[nb], likerettar[nn] (nb) njuolggan (se) likriktare (sv) tasasuuntaaja (fi)

reduction valve

reduksjonsventil (nb) geahpidanventiila (se) tryckreduceringsventil (sv) paineenalennusventtiili (fi)

reel

See "coil, spool, reel".

reel, swift, capstan

haspel (nb) vikša (se) haspel (sv) kela, vintturi (fi)

register

register (nb) registtar (se) register (sv) rekisteri (fi)

register belt, timing belt

registerreim (nb) registtarruopma (se) kuggrem, registerrem, kamaxelrem (sv) jakopään hihna, jakohihna (fi)

register chain, timing chain

registerkjede (nb) registtarviđji (se) registerkedja, kamaxelkedja (sv) jakopään ketju, jakoketju (fi)

regulation

See "governing, regulation, adjustment".

regulator, governor

regulator (nb) reguláhtor (se) regulator (sv) säädin, säätäjä (fi)

reinforce

See "strengthen, reinforce".

reinforcement

See "strengthening, reinforcement".

reinforcement steel

armeringsstål, kamstål (nb) nannenstálli, nanosmahttinstálli (se) armeringsstål (sv) betoniteräs (fi)

reinforcing

See "armouring, reinforcing".

relay

rele (nb) rele (se) relä (sv) rele (fi)

relief angle

See "clearance angle, relief angle".

remote control

fjernstyring, fjernkontroll (nb) gáiddusstivren (se) fjärrmanövrering (sv) kaukoohjaus (fi)

remove burrs

See "clean off burrs, remove burrs".

repair

reparere (nb) divvut (se) reparera (sv) korjata (fi)

repair, restore, mend

reparasjon (nb) divvun (se) reparation (sv) korjaus (fi)

reservoir

See "container, bin, reservoir, tank".

resistance

motstand, resistans (nb) vuosttaldus, resistánsa (se) motstånd (sv) vastus (fi)

resistance power

motstandsevne (nb) vuostenákca (se) motståndskraft (sv) vastavoima, vastustuskyky (fi)

restore

See "repair, restore, mend".

return line

returledning[nb], returleidning[nn] (nb) máhccanjohdas (se) returledning (sv) paluujohto (fi)

reusable pallet

palle, pall (nb) palla (se) pall (sv) trukkilava, jakkara (fi)

reverse

revers (nb) murdinmolsson (se) backväxel (sv) peruutusvaihde (fi)

revolution

omdreiing (nb) jorran (se) varv (sv) kierto (fi)

revolution counter, tachometer

omdreiingsmåler[nb], -målar[nn], turteller[nb], -teljar[nn] (nb) jorranmihtádas (se) varvmätare (sv) kierroslukumittari, pyörimisnopeusmittari (fi)

revolving punch pliers

See "punch, revolving punch pliers".

rib, fin, web, gill

ribbe (nb) erttet (se) spant, utsprång, ribba (sv) rima, ripa (fi)

rib, frame

spant (nb) fierra, rággu (se) spant (sv) kaari, laita (fi)

rifle

See "serrate, rifle, flute, groove".

rightward welding

motsveising, høyresveising[nb], høgresveising[nn] (nb) vuostesveisen (se) motsvetsning (sv) vastahitsaus (fi)

rigid ball bearing

See "track roller bearing, rigid ball bearing".

rigidity

See "stiffness, rigidity".

rim, felly, felloe

felg (nb) fealga (se) fälg (sv) vanne (fi)

rinsing nozzle

See "washing nozzle, rinsing nozzle".

ripple

See "knurl, flute, groove, ripple".

rivet

klinke (nb) duorrat (se) nita (sv) niitata (fi)

rivet

nagle, klinknagle (nb) návli, duorrannávli (se) nit (sv) niitti (fi)

rivet tool, hand riveter

popnagletangb blindnagletang[nb], -tong[nn] (nb) popnávledoaŋggat (se) nittång (sv) pop-niittipihdit (fi)

rod, hanger, strutting

stag (nb) snáhki (se) stag (sv) side, tuki (fi)

roll

See "cylinder, roll, roller, barrel".

roller

See "cylinder, roll, roller, barrel".

roller

rull (nb) rulla (se) rulle (sv) rulla (fi)

roller bearing

rullelager (nb) rullaláger (se) rullager (sv) rullalaakeri (fi)

rolling bearing

rullingslager (nb) jorreláger (se) rullningslager (sv) vierintälaakeri (fi)

rolling machine

valsemaskin, valse (nb) ludnemašiidna, válsamašiidna (se) valsmaskin (sv) valssikone (fi)

rolling mill

valseverk (nb) válsadoaimmahat (se) valsverk (sv) valssilaitos, valssaamo, valssain (fi)

rotate

rotere (nb) jorrat (se) rotera (sv) pyöriä (fi)

rotational speed, rpm

omdreiingstall, turtall, -tal[nn] (nb) jorranlohku (se) varvtal (sv) pyörimisnopeus (fi)

rotor

rotor, anker (nb) rohtor, jorri (se) rotor (sv) roottori, ankkuri (fi)

rough edge

See "burr, rough edge".

rough-finish, roughing

grovbearbeiding (nb) roavvagieđahallan (se) grovbearbetning (sv) rouhinta, karkeatyöstö (fi)

roughing

See "rough-finish, roughing".

roughing tool

skrubbstål (nb) roavvastálli (se) skrubbstål (sv) rouhintaterä (fi)

roughness

ruhet[nb], ruleik[nn] (nb) romšodat, roamši (se) råhet (sv) karkeus (fi)

round end pliers

rundtang[nb], rundtong[nn] (nb) jorbabasttat (se) rundtång (sv) pyörökärkipihdit (fi)

round file

rundfil (nb) jorbafiilu (se) rundfil (sv) pyöröviila (fi)

rowel, pulley, caster

trinse (nb) jorregeavri (se) trissa (sv) väkipyörä (fi)

rpm

See "rotational speed, rpm".

rubber

gummi (nb) gummi (se) gummi (sv) kumi (fi)

rubbing

gniding (nb) ruvven (se) gnidning (sv) hankaaminen (fi)

rule

rettholt, reie (nb) njuolggolinjála (se) rätskiva (sv)

ruler

linjal (nb) linjála (se) linjal (sv) viivain, viivoitin (fi)

running

See "operation, running".

run-out

See "eccentricity, run-out".

rupture disc, spring washer

sprengskive, fjærskive[nb], fjørskive[nn] (nb) dávgerovvi (se) fjäderbricka (sv) jousialuslaatta, jousialuslevy (fi)

rust

rust (nb) ruosta (se) rost (sv) ruoste (fi)

rust, corrode

ruste (nb) ruostut (se) rosta (sv) ruostua (fi)

rust killer, rust removing fluid

rustløser[nb], rustløysar[nn] (nb) ruostaluvvi (se) rostlösare (sv) ruosteenirroitin (fi)

rust removing fluid

See "rust killer, rust removing fluid".

safe-edge file

See "flat file, blunt file, safe-edge file".

safety trustee

verneombud[nb], verneombod[nn] (nb) suodjalusáittardeaddji (se) skyddsombud (sv) työsuojeluasiamies (fi)

sandblasting

sandblåsing (nb) sáddobossun (se) sandblästring (sv) hiekkapuhallus (fi)

sandpaper

See "emery paper, sandpaper".

saturate

mette (nb) gallehit (se) mätta (sv) kyllästää (fi)

save, record

lagre (nb) vurket (se) lagra (sv) tallentaa (fi)

saw

sag (nb) sahá (se) såg (sv) saha (fi)

saw blade

sagblad (nb) sahádearri (se) sågblad (sv) sahanterä (fi)

scale, oxide scale

glødeskall[nb], glødeskal[nn] (nb) áhcagastingarra (se) glödskal (sv) hehkutushilse, hilseyly (fi)

scale, table

skala (nb) skála (se) skala (sv) asteikko, mittakaava (fi)

scanner

bildeleser[nb], biletlesar[nn], skanner (nb) govvalohkki (se) avsökare, scanner (sv) kuvanlukija (fi)

scarf

See "rabbet, scarf".

scissors, shears

saks (nb) skárrit, skierat (se) sax (sv) sakset, leikkuri (fi)

scoop

See "blade, scoop, bucket, vane".

scouring machine, washing machine

delevasker[nb], delevaskar[nn] (nb) oasselávgu (se) osienpesulaite (fi)

scrape

skrape (nb) faskut (se) skrapa (sv) kaapia (fi)

scraper

skrape (nb) faskkon (se) skrapa (sv) kaavin (fi)

scrap iron

skrapjern (nb) šlámbarruovdi, roaisteruovdi (se) järnskrot (sv) romurauta (fi)

scratch awl

See "scribing awl, scriber, scratch awl".

scratching

riss (nb) sáhcu (se) rits (sv) piirrotus (fi)

scratch, scribe

risse, ripe (nb) sáhcut (se) ritsa (sv) piirrottaa (fi)

screen, monitor, display

skjerm (nb) šearbma (se) bildskärm (sv) näyttö (fi)

screw

skrue (nb) skruvva (se) skruv (sv) ruuvi (fi)

screw clamp

See "clamp frame, screw clamp".

screw die

See "cutting die, threading die, screw die".

screwdriver

skrujern, skrutrekker[nb], skrutrekkjar[nn] (nb) skruvvaruovdi, skruvvabonjan (se) skruvmejsel (sv) ruuvitaltta, ruuvimeisseli (fi)

screw extractor

skrueuttrekker[nb], skrueuttrekkjar[nn], grisepikk (nb) skruvvageasán (se) skruvutdragare (sv) ruuvin ulosvetäjä (fi)

screw feeder

See "feed worm, screw feeder".

screw pitch gauge

See "thread gauge, screw pitch gauge".

screw pitch, pitch of thread

gjengestigning, stigning (nb) jeaŋgagoargŋun (se) gängstigning (sv) (kierteen) nousu (fi)

screw tap

See "tap, screw tap, thread tap".

scribe

See "scratch, scribe".

scriber

See "scribing awl, scriber, scratch awl".

scribing awl, scriber, scratch awl

rissenål (nb) sáhcunnállu, soairu (se) ritsnål (sv) piirtopuikko, piirtokärki (fi)

scroll, spiral groove

spiralspor (nb) bonjisgurra (se) spiralspår (sv) kierreura (fi)

sealing device

See "seal, seat, sealing device, packing".

sealing face

See "contact face, sealing face".

sealing ring

tetningsring[nb], tettingsring, gacoring (nb) lávgadangierdu, lávgadanriekkis (se) tätningsbricka (sv) tiivisterengas (fi)

seal, seat, sealing device, packing

tetning[nb], tetting (nb) lávgadas (se) tätning (sv) tiiviste, tiivistin (fi)

seam

See "fold, flange, seam, joint".

seat

See "seal, seat, sealing device, packing".

seat

sete (nb) čohkkedat (se) säte (sv) istuin (fi)

secondary

sekundær (nb) nuppáldas, sekundára (se) sekundär (sv) toisio- (fi)

secondary edge, minor cutting edge

biegg (nb) oalgeávju (se) biskäregg, biegg (sv) sivusärmä (fi)

second tap, plug tap

mellomtapp, midttapp (nb) gaskadáhppa (se) mellantapp (sv) keskimmäinen kierretappi (fi)

section

See "cut, notch, section".

securing plate

låseblikk[nb], låseblekk[nn] (nb) lássenbelle, -spelle, lohkkadan- (se) låsbleck (sv) lukituslevy, varmistuslevy (fi)

self-centering

sjølsentrerende[nb], sjølvsentrerande[nn] (nb) iešguovddášahtti (se) självcentrerande (sv) itsekeskittävä (fi)

self-locking screw

sjølsperrende skrue[nb], sjøllåsende skrue[nb], sjølvlåsande skrue[nn] (nb) iešcaggi skruvva, iešdoalli skruvva (se) självhämmande skruv (sv) itsepidättävä ruuvi (fi)

self-tapping screw

sjølgjengende skrue[nb], sjølvgjengande skrue[nn] (nb) iešjeŋgenskruvva (se) gängpressande skruv (sv) itsekierteittävä ruuvi (fi)

semi-circular protractor

vinkeltransportør (nb) čiehkasirddan (se) vinkeltransportör (sv) yleiskulmamitta (fi)

semiconductor

halvleder[nb], halvleiar[nn] (nb) beallejođadas (se) halvledare (sv) puolijohde (fi)

semi-products

See "half finished goods, semi-products".

sensor

See "detector, sensor, probe".

separator

See "insulator, separator".

serial connection

See "series connection, serial connection".

seriel interface

serielt grensesnitt (nb) maŋŋálaslakta (se) seriellt gränssnitt (sv) sarjaliitäntä (fi)

series connection, serial connection

seriekopling (nb) ráidogoallus (se) seriekoppling (sv) sarjakytkentä (fi)

serrate, rifle, flute, groove

serratere (nb) rihkkuhit, serrateret (se) räffla (sv) pyältää, rihloittaa, urittaa (fi)

serrate tool

serrat (nb) rihkon, serrat (se) pyällyskehrä, pyällystyökalu (fi)

service

See "maintenance, service".

set

vikke (nb) hirret, botnjat, vihkket (se) skränka (sv) harittaa (fi)

set screw

See "adjusting screw, set screw".

set screw, fixing screw

settskrue, festeskrue (nb) giddenskruvva (se) stoppskruv, fästskruv (sv) kiinnitysruuvi, pidätinruuvi (fi)

setting angle, cutting edge angle

innstillingsvinkel (nb) čuohppanávjučiehka (se) ställvinkel (sv) asetuskulma (fi)

sewer pipe

See "drain pipe, outlet tube, sewer pipe".

shaft

See "haft, shaft, handle, grip".

shaft, axle

aksel, aksling (nb) áksil (se) axel (sv) akseli (fi)

shaft extension

See "shaft journal, shaft extension".

shaft journal, shaft extension

akseltapp (nb) aksilculci, aksildáhppa (se) axeltapp (sv) akselitappi (fi)

shale, waste rock

gråberg (nb) ránesbákti (se) gråberg (sv) jätekivi, hylkykivi (fi)

shaper

See "planer, shaper, planing machine, shaping machine".

shaping machine

See "planer, shaper, planing machine, shaping machine".

sharpen

See "grind, sharpen".

sharpening

See "grinding, sharpening".

shavings

See "turnings, shavings".

shears

See "scissors, shears".

sheath knife

tollekniv (nb) buiku, buvku (se) slidkniv, täljkniv (sv) puukko (fi)

sheet

See "plate, sheet".

sheet metal shears, plate shears

platesaks (nb) pláhttaskárrit, pláhttaskierat (se) plåtsax (sv) levysakset (fi)

sheet metal, tin plate

blikk[nb], blekk[nn] (nb) belle, spelle (se) bleck, plåt (sv) pelti (fi)

shell end mill, end face mill

endeplanfres (nb) geahčedulbenjurssan (se) ändplanfres (sv) lieriöotsajyrsin (fi)

shield gas

dekkgass (nb) suodjegássa (se) skyddsgas (sv) suojakaasu (fi)

shift

skift (nb) vuorru (se) skift (sv) vuoro (fi)

shift gear

See "change gear, shift gear, alternate".

shim

mellomlegg, skims (nb) gaskkus (se) mellanlägg (sv) täytelevy, sovituslevy (fi)

shock absorber

støtdemper[nb], støytdempar[nn] (nb) haŋkinváiddon, časkinváiddon (se) stötdämpare (sv) heilahduksen vaimennin, iskunvaimennin (fi)

short-circuit

kortslutning (nb) njuolggogidden, njuolggolaktin (se) kortslutning (sv) oikosulku (fi)

shoulder, collar, lug

ansats, krage (nb) stealli (se) ansats (sv) olake, ulkonema (fi)

shrink

krympe (nb) duohpahuvvat, duohpašuvvat (se) krympa (sv) kutistua (fi)

shrinkage

See "shrinking, shrinkage".

shrinking, shrinkage

krymping (nb) duhppen (se) krympning (sv) kutistuminen, kutistuma (fi)

shrink on, crimp

krympe på (nb) duhppet (se) krympa på (sv) asentaa kutistamalla (fi)

shunt, shunt switch, parallel connection

parallellkopling (nb) paralleallagoallus, buohtalatgoallus (se) parallellkoppling (sv) rinnankytkentä, rinnakkaiskytkentä (fi)

shunt switch

See "shunt, shunt switch, parallel connection".

side cutting pliers

sideavbitertang[nb], sideavbitartong[nn], skråavbiter (nb) gilgacikcenbasttat, vinjubasttat (se) sidavbitare (sv) sivuleikkurit (fi)

side of thread, flank of thread

gjengeflanke (nb) jeaŋgagilga (se) gängflank (sv) kierteen kylki (fi)

sieve

See "strainer, sieve".

signal

signal (nb) mearka, signála (se) signal (sv) merkki, signaali (fi)

signal horn

See "horn, signal horn".

silencer, damper, muffler, exhaust box

eksospotte, lyddemper[nb], lyddempar[nn] (nb) bázahasbohttu, eksosbohttu, jietnaváiddon (se) ljuddämpare (sv) äänenvaimennin (fi)

silencer, muffler, damper

lyddemper[nb], lyddempar[nn] (nb) jietnaváiddon (se) ljuddämpare (sv) äänenvaimennin (fi)

single-acting cylinder

enkeltvirkende sylinder[nb], enkeltverkande sylinder[nn] (nb) ovttadoaimmat sylinddar (se) enkelverkande cylinder (sv) yksitoiminen sylinteri (fi)

sintering

sintring (nb) sintren (se) sintring (sv) sintraus (fi)

site

See "situation, site, location".

situation, site, location

beliggenhet[nb], posisjon[nn] (nb) sajádat (se) läge (sv) sijainti, paikka, asento (fi)

skid

See "slip, skid, slide".

slag, cinder

slagg (nb) buollángarra, bázahasgarra (se) slagg (sv) kuona (fi)

slag hammer

slagghammer[nb], slagghammar[nn], slaggpikke (nb) bázahasveažir (se) slagghammare (sv) kuonavasara, kuonahakku (fi)

sledge hammer, forging hammer

smihammer[nb], smihammar[nn] (nb) bádjeveažir (se) skenhammare (sv) pajavasara (fi)

sledge hammer, mallet, maul

slegge, sleggje[nn] (nb) šleagga (se) slägga (sv) moukari, leka (fi)

sleeve

See "bush(ing), sleeve, socket".

sleeve, collar

mansjett (nb) manšeahtta (se) skyddshylsa (sv) suojaholkki, kaulus (fi)

slide

gli (nb) johtit (se) glida (sv) liukua (fi)

slide

sleide (nb) mielggas (se) slid (sv) kelkka, luisti (fi)

slide

See "slip, skid, slide".

slide coupling, slipping clutch

slurekopling (nb) njalpalavtta, njalpagoallus (se) slirkoppling (sv) liukukytkin (fi)

slide fit

skyvepasning (nb) hoiganheivehus (se) skjutpassning (sv) työntötiukkuus (fi)

slide gauge

See "sliding caliper, vernier caliper, slide gauge".

slide valve

See "gate valve, slide valve".

sliding bearing, plain bearing

glidelager (nb) njoalveláger, johtinláger (se) glidlager (sv) liukulaakeri (fi)

sliding caliper, vernier caliper, slide gauge

skyvelære, skyvemål (nb) sirdinmihtádas, hoiganmihtádas (se) skjutmått (sv) työntömitta (fi)

sliding friction

See "sliding resistance, sliding friction".

sliding resistance, sliding friction

glidemotstand (nb) johtinvuosti (se) glidmotstånd (sv) liukuvastus (fi)

slipping clutch

See "slide coupling, slipping clutch".

slip, skid, slide

slure (nb) njalpit (se) slira (sv) luistaa, luistattaa, liukua (fi)

slot, groove

spor (nb) gurra, luodda (se) spår (sv) ura (fi)

slotted screw

sporskrue (nb) gurraskruvva (se) spårskruv (sv) uraruuvi (fi)

slotting tool

stikkstål (nb) čuggenstálli (se) stickstål (sv) pistoterä, pistin (fi)

smith, blacksmith, forger

smed (nb) rávdi (se) smed (sv) seppä (fi)

smithy, forge

smie (nb) bádji (se) smedja (sv) paja, työpaja (fi)

snap ring

sikringsring, orepinne (nb) sihkkarastinriekkis (se) säkringsring (sv) lukkorengas (fi)

snow scooter

snøskuter, beltemotorsykkel (nb) mohtorgielka, skohter (se) snöskoter (sv) moottorikelkka, skootteri (fi)

socket

See "box, socket".

socket, contact

stikkontakt, kontakt (nb) el-čuggestat, ráigeguoskkahat (se) vägguttag, väggkontakt, kontakt (sv) pistorasia (fi)

socket, sleeve, faucet, collar

muffe (nb) muffa (se) muff (sv) holkki, muhvi (fi)

socket wrench

See "box wrench, socket wrench".

soft soldering, tin soldering

tinnlodding, mjuklodding, bløtlodding[nb] (nb) datnejugaheapmi (se) tennlödning (sv) tinajuotos (fi)

solar cell

solcelle (nb) beaivvášsealla (se) solcell (sv) aurinkokenno (fi)

solder

lodde (nb) jugahit (se) löda (sv) juottaa (fi)

soldering

lodding (nb) jugaheapmi, jugahahttin (se) lödning (sv) juottaminen (fi)

soldering iron

loddebolt (nb) golve (se) lödkolv (sv) juotin, juottokolvi (fi)

soldering paste

See "flux, soldering paste, fluidizer".

soldering tin

loddetinn (nb) jugahandatni (se) lödtenn (sv) juottotina (fi)

solid

massiv (nb) dievaslaš (se) massiv (sv) täyteinen (fi)

solution

See "emulsion, solution".

solvent

løsemiddel[nb], løysemiddel (nb) luvvadanávnnas, luvvadat (se) lösningsmedel (sv) liuote, liuotinaine (fi)

soot

sote (nb) giebahuhttit, gožuhuhttit (se) sota (sv) noeta (fi)

soot

sot (nb) giehpa, gožu (se) sot (sv) noki (fi)

sound testing

klangprøve (nb) šuokŋaiskan (se) klangkontroll (sv) koputuskoe (fi)

source of current

See "power source, source of current".

spanner, box wrench, socket wrench

stjernenøkkel, ringnøkkel (nb) nástečoavdda, gierdočoavdda (se) ringnyckel (sv) lenkkiavain, silmukka-avain, rengasavain (fi)

spanner, wrench

skrunøkkel (nb) skruvvenčoavdda (se) skruvnyckel (sv) ruuviavain (fi)

span of jaws

See "width across flats, span of jaws".

spare part

reservedel (nb) liigeoassi (se) reservdel (sv) varaosa (fi)

spark

gnist[nb], gneiste[nn] (nb) čuonan (se) gnista (sv) kipinä (fi)

spark coil

See "ignition coil, spark coil".

spark plug cable

See "plugwire, spark plug cable".

spark plug cap

plugghette (nb) ginttalgahpir (se) sytytystulpan hattu (fi)

spark plug, ignition plug

tennplugg (nb) cahkkehanginttal (se) tändstift (sv) sytytystulppa (fi)

spark plug wrench

pluggnøkkel, tennpluggnøkkel (nb) pluggačoavdda, ginttalčoavdda (se) tändstiftsnyckel (sv) tulppa-avain (fi)

spattle, spatula, filling knife

sparkel (nb) sparkel, deavdinniibi (se) spackel (sv) lasta (fi)

spatula

See "spattle, spatula, filling knife".

speedometer

hastighetsmåler[nb], fartsmåler[n fartsmålar[nn] (nb) leaktomihtádas (se) hastighetsmätare (sv) nopeusmittari (fi)

speed, rate, velocity

hastighet[nb], fart (nb) leaktu, leahttu (se) hastighet (sv) nopeus, vauhti (fi)

spesific

spesifikk (nb) ieš- (se) specifik (sv) ominais- (fi)

spherical

sfærisk (nb) sfearalaš (se) sfärisk (sv) pallomainen (fi)

spherical bearing

sfærisk lager (nb) sfearalaš láger (se) sfäriskt lager (sv) pallomainen laakeri (fi)

spindle

See "head stock, mandrel stock, spindle".

spindle, stem

spindel (nb) jorreáksil (se) spindel (sv) kara (fi)

spiral drill

See "twist drill, spiral drill".

spiral groove

See "scroll, spiral groove".

splines

flerkileforbindelse[nb], fleirkilesamband[nn],

splines (nb) máŋggalođatovttastus (se) bomspår, splines (sv) akselin uritus, rihlaus (fi)

spline shaper, broach

kilsporhøvel, trekkbrosj (nb) lohtegurraheavval, lohteluoddaheavval (se) kilspårshyvel, kilspårfräs, dragbrotsch (sv) kiilauranhöylä (fi)

splinter

See "splint, splinter, pin, cotter, split pin".

splint, splinter, pin, cotter, split pin

splint, splittpinne, saksesplint (nb) suorresággi (se) sprint, saxpinne (sv) sokkanaula (fi)

split pin

See "splint, splinter, pin, cotter, split pin".

spool

See "coil, spool, reel".

spot welder, spot welding machine

punktsveiseapparat (nb) čuokkissveisenapparáhta (se) punktsvetsningsmaskin (sv) pistehitsauskone (fi)

spot welding

punktsveising (nb) čuokkissveisen (se) punktsvetsning (sv) pistehitsaus (fi)

spot welding machine

See "spot welder, spot welding machine".

spray nozzle

See "nozzle, spray nozzle, jet".

spring

fjær[nb], fjør[nn] (nb) dávgi (se) fjäder (sv) jousi (fi)

spring steel

fjærstål[nb], fjørstål[nn] (nb) dávgestálli (se) fjäderstål (sv) jousiteräs (fi)

spring tension

fjærspenning[nb], fjørspenning[nn] (nb) dávgegealdda (se) fjäderspänning (sv) jousenjännitys (fi)

spring washer

See "rupture disc, spring washer".

square, angle gauge, bevel protractor

vinkel (nb) viŋkil, čiehka (se) vinkel (sv) suorakulma (fi)

square head screw

firkantskrue (nb) njealječiegat skruvva, njealjeborat skruvva (se) fyrkantskruv (sv) neliökantaruuvi (fi)

square, joint hook

platevinkel (nb) pláhttaviŋkil (se) flackvinkel (sv) suorakulma (fi)

square thread

See "flat thread, square thread".

square, try square, joint hook

anslagsvinkel, ansatsvinkel, vinkelhake (nb) stealleviŋkil, steallečiehka (se) vinkelhake (sv) kulmaviivain, suorakulmain (fi)

stability

stabilitet (nb) stáđisvuohta, stargatvuohta (se) stabilitet (sv) vakavuus, pysyvyys (fi)

stable, steady

stabil (nb) stáđis, starga (se) stabil (sv) pysyvä, vakaa (fi)

stainless steel

rustfritt stål (nb) ruostameahttun stálli (se) rostfritt stål (sv) ruostumaton teräs (fi)

stake

See "anvil, stake".

stamp

See "punch, stamp".

stamp(er)

See "punching tool, stamp(er)".

stamping machine

See "punching machine, stamping machine".

stand, rack, frame

stativ (nb) holga (se) stativ (sv) jalusta, runko, kehys (fi)

star-delta connection

stjernetrekantkopling (nb) nástegolbmačiegatgoallus (se) stjärntriangelkoppling (sv) tähtikolmiokytkentä (fi)

start

starte (bil, motor) (nb) vuolggahit (se) starta (sv) käynnistää (fi)

starter

selvstarter[nb], sjølvstartar[nn] (nb) čoavddavuolggahus (se) käynnistin (fi)

starter

See "starting motor, starter".

starter gear ring

startkrans (nb) vuolggahangierdu (se) startkrans (sv) käynnistyshammaskehä (fi)

starting motor, starter

startmotor (nb) vuolggahanmohtor (se) startmotor (sv) käynnistysmoottori (fi)

starting point

See "modular measure, starting point".

static

statisk (nb) stahtalaš (se) statisk (sv) staattinen, tasapainoinen (fi)

stator

stator (nb) stahtor (se) stator (sv) staattori (fi)

steady

See “stable, steady”.

steam

damp (nb) lievla, levlo (se) ånga (sv) höyry (fi)

steam engine

dampmaskin (nb) lievlamašiidna, levlomašiidna (se) ångmaskin (sv) höyrykone (fi)

steel

stål (nb) stálli (se) stål (sv) teräs (fi)

steel shavings

See “steel wool, steel shavings”.

steel tape rule

See “measuring tape, steel tape rule”.

steel wool, steel shavings

stålull (nb) stálleullu (se) stålull (sv) teräsvilla (fi)

steering, control, operation

styring (nb) stivren (se) styrning (sv) ohjaus (fi)

steering wheel

See “hand wheel, steering wheel”.

stem

See “spindle, stem”.

stepless

trinnløs[nb], trinnlaus (nb) ceahkeheapmi (se) steglös (sv) portaaton (fi)

stiffness, rigidity

stivhet[nb], stivleik[nn] (nb) stargatvuohta, stargodat, stirddisvuohta (se) styvhet (sv) jäykkyys (fi)

stirrup

See “bow, clamp, stirrup, loop”.

stock

See “haft, stock”.

straightening

See “alignment, straightening”.

straightening tool

opprettingsverktøy (nb) njulgenneavvu (se) riktverktyg (sv) oikaisutyökalut (fi)

straight pane hammer, pin hammer

pennhammer[nb], pennhammar[nn] (nb) peannaveažir, duorranveažir (se) penhammare (sv) harjapäävasara (fi)

strain

See “percolate, filter, strain”.

strainer

See “filter, strainer”.

strainer, sieve

sil (nb) silli (se) sil (sv) siivilä, suodatin (fi)

strap

See “driving belt, strap, band”.

strength

fasthet[nb], fastleik[nn] (nb) nannodat (se) fasthet (sv) kestävyys, vahvuus (fi)

strengthening, reinforcement

forsterkning[nb], forsterking (nb) gievrudus (se) förstärkning (sv) vahvistus, vahvistaminen (fi)

strengthen, reinforce

forsterke (nb) gievrudit, nanosmahttit (se) förstärka (sv) vahvistaa (fi)

stress

See "tension, stress".

stripping pliers

avisoleringstang[nb], avisoleringstong[nn] (nb) garahanbasttat (se) skaltång (sv) kuorintapihdit (fi)

stripping pliers

kabelskotang, kabelskotong[nn] (nb) jođasgámabasttat (se) kabelskotång (sv) kaapelikenkäpihdit (fi)

stroke, piston stroke

takt, stempelslag (nb) dákta (se) takt, kolvslag (sv) tahti, männän isku (fi)

structural steel

See "construction steel, structural steel".

strutting

See "rod, hanger, strutting".

stud

See "peg, pin, plug, stud".

stud torque

tiltrekkingsmoment (nb) čavgenmomeanta (se) kiristysmomentti (fi)

suction, inlet

innsuging, sug (nb) njamman (se) sug (sv) imu (fi)

sulphur

svovel (nb) rišša (se) svavel (sv) rikki (fi)

support

anlegg (nb) stealli (se) stöd (sv) tuki (fi)

surface

overflate (nb) olggoš (se) yta (sv) pinta (fi)

surface cutter

See "facing cutter, surface cutter".

surface gauge, marking gauge

høyderisser[nb], høgderissar[nn], rissefot (nb) allodatsárggon (se) ritskubb (sv) piirrotuskelkka, suuntapiirrin (fi)

surface grinding machine

planslipemaskin (nb) dulbenšliipenmašiidna (se) planslipmaskin (sv) tasohiomakone (fi)

surface norm, surface rule

overflatenormal (nb) olggošmálle (se) ytnormal (sv) pinnankarheusmalli (fi)

surface of fracture

bruddflate[nb], brotflate[nn] (nb) doddjonolggoš (se) brottyta (sv) murtopinta (fi)

surface plate

planbord (nb) duolbabeavdi (se) riktplatta (sv) oikotaso, väritaso (fi)

surface roughness

overflateruhet[nb], overflateruleik[nn] (nb) olggošromšodat, olggošroamši (se) ytråhet (sv) pinnankarheus (fi)

surface rule

See "surface norm, surface rule".

surface treatment

overflatebehandling (nb) olggošgieđahallan (se) ytbehandling (sv) pintakäsittely (fi)

surfacing

See "coat, coating, facing, lining, surfacing".

swarf

bruddspon[nb], brotspon[nn] (nb) doddjonvuolahas (se) kortspån (sv) katkolastu, murtolastu (fi)

swift

See "reel, swift, capstan".

switch, circuit breaker, interrupter

bryter[nb], brytar[nn] (nb) botkkon (se) brytare (sv) kytkin, katkaisin (fi)

swivel handle, swivel rod

kraftarm, leddhandtak (nb) fápmonađđa (se) ledhandtag (sv) nivelväännin (fi)

swivel rod

See "swivel handle, swivel rod".

swivel socket wrench

leddnøkkel (nb) lađasčoavdda (se) ledhylsnyckel (sv) nivelavain (fi)

symmetric

symmetrisk (nb) symmetralaš (se) symmetrisk (sv) symmetrinen (fi)

synchronous

synkron (nb) synkruvdnalaš (se) synkron (sv) synkroninen, tahdistettu (fi)

synchronous motor

synkronmotor (nb) synkronmohtor (se) synkronmotor (sv) tahtimoottori, synkronimoottori (fi)

synthetic

syntetisk (nb) syntetalaš (se) syntetisk (sv) synteettinen (fi)

system

system (nb) vuogádat, systema (se) system (sv) järjestelmä, systeemi (fi)

table

See “scale, table”.

tachometer

See “revolution counter, tachometer”.

tackle, hoist pulley

talje (nb) dáhkkal (se) talja, blocktyg (sv) talja (fi)

tail stock

bakdokke, pinoldokke (nb) geahčedoalan (se) dubbdocka, pinoldocka (sv) kärkipylkkä (fi)

tail stock centre

senterspiss, pinolspiss (nb) guovddášnjunni, pinolnjunni (se) dubb (sv) keskiökärki (fi)

tail stock spindle

pinolrør, pinol (nb) pinolbohcci (se) dubbrör, pinol (sv) siirtopylkän holkki (fi)

tang, cone

tange (nb) bortnis (se) tånge (sv) ruoto (fi)

tank

See “container, bin, reservoir, tank”.

tap

See “cock, tap, faucet”.

taper clamping sleeve

spennhylse (nb) čávgenskuohppu (se) spännhylsa (sv) kiristysholkki (fi)

taper, cone

kon (nb) lávvolat, čohkolaš (se) kona (sv) kartio (fi)

tapered

See “conical, coned, tapered, coniform”.

tapered sleeve, drill coupling

borhylse (nb) bovraskuohppu (se) borrchuck (sv) väliholkki, supistusholkki (fi)

taper pin

konpinne (nb) čohkkosággi (se) konpinne (sv) kartiotappi (fi)

taper tap, pilot tap

spisstapp, forgjengetapp (nb) álgodáhppa (se) förtapp (sv) aloituskierretappi, esikierretappi (fi)

taper turning

kondreiing, konisk dreiing (nb) lávvolatvárven, čohkolašvárven (se) konsvarvning (sv) kartiosorvaus (fi)

tap, screw tap, thread tap

gjengetapp (nb) jeŋgendáhppa (se) gängtapp (sv) kierretappi (fi)

telescopic gauge

teleskopmål (nb) teleskohpamihtádas (se) teleskopstickmått (sv) pistomitta (fi)

temper

See "harden, temper".

tempered steel

seigherda stål (nb) dávggasbuoššoduvvon stálli (se) seghärdat stål (sv) nuorrutettu teräs, päästökarkaistu teräs (fi)

tempering

See "hardening, tempering".

tempering

adusering (nb) aduseren (se) aducering (sv) adusointi (fi)

tempering

seigherding (nb) dávggasbuoššodeapmi (se) seghärdning (sv) nuorrutus, päästökarkaisu (fi)

tempering furnace

See "hardening furnace, tempering furnace".

template

See "gauge, jig, template".

tenon saw, compass saw, dowel saw

stikksag, rotterumpe (nb) buvkosahá, buškosahá (se) sticksåg (sv) pistosaha (fi)

tensile strength

strekkfasthet[nb], strekkfastleik[nn (nb) fatnannannodat (se) draghållfasthet (sv) vetolujuus (fi)

tension, stress

spenning (nb) gealddа (se) spänning (sv) jännitys (fi)

test

prøve, test (nb) iskan (se) prov (sv) koe (fi)

T- handle

T-arm, T-handtak (nb) T-nađđa (se) T-handtag (sv) T-väännin (fi)

thinner

tynner[nb], tynnar[nn] (nb) njárbbadas (se) förtunningsmedel (sv) ohennin (fi)

thread

gjenge (nb) jeaŋga (se) gänga (sv) kierre (fi)

thread, cut threads

gjenge (nb) jeŋget (se) gänga (sv) kierteittää (fi)

thread gauge, screw pitch gauge

gjengelære, gjengesøker[nb], gjengesøkjar[nn] (nb) jeaŋgamihtádas (se) gängtolk (sv) kierremitta, kierrekampa (fi)

threading die

See "cutting die, threading die, screw die".

threading tool

See "chaser, chasing tool,threading tool".

thread tap

See "tap, screw tap, thread tap".

three phase motor

trefasemotor (nb) golmmamuddutmohtor (se) trefasmotor (sv) kolmivaihemoottori (fi)

throttle valve, choke

strupeventil (nb) buvihanventiila (se) strypventil (sv) virtavastusventtiili (fi)

thrust

See "motive power, thrust, motive force".

thrust bearing

aksiallager (nb) aksiálláger (se) axiallager, trycklager (sv) aksiaalilaakeri, päittäislaakeri (fi)

thyristor

tyristor (nb) tyristor (se) tyristor (sv) tyristori (fi)

tighten

See "fasten, tighten, fix".

tight fit, drive fit

drivpasning (nb) čavgaheivehus (se) drivpassning (sv) pakotustiukkuus (fi)

timing belt

See "register belt, timing belt".

timing chain

See "register chain, timing chain".

tin

tinn (nb) datni (se) tenn (sv) tina (fi)

tinman

See "tinsmith, tinman".

tin plate

See "sheet metal, tin plate".

tin shears, tinsnips

blikksaks[nb], blekksaks[nn] (nb) belleskárrit, -skierat, spelle- (se) plåtsax, blecksax (sv) peltisakset, levysakset (fi)

tinsmith, tinman

blikkenslager[nb], blekkslagar[nn] (nb) bellečeahppi, bellerávdi, spelle- (se) bleckslagare, plåtslagare (sv) levyseppä, peltiseppä (fi)

tinsnips

See "tin shears, tinsnips".

tin soldering

See "soft soldering, tin soldering".

title block

tittelfelt (nb) nammaoassi (se)

tolerance, allowance, margin

toleranse (nb) toleránsa (se) tolerans (sv) toleranssi (fi)

tolerance of fit

See "fit, tolerance of fit".

tolerance range

toleranseområde (nb) toleránsaguovlu (se) toleransområde (sv) toleranssialue (fi)

tongs, pliers

tang[nb], tong[nn] (nb) basttat, doaŋggat (se) tång (sv) pihdit (fi)

tool

verktøy, verkty[nn] (nb) bargoneavvu, reaidu (se) verktyg (sv) työkalu (fi)

tool box

verktøykasse, verktøykiste (nb) neavvobumbá, bargobumbá (se) verktygslåda (sv) työkalulaatikko, työkalupakki (fi)

tool cutting edge inclination

See "rake, tool cutting edge inclination".

tool holder

verktøyfeste (nb) neavvogiddehat (se) verktygshållare (sv) teränpidin (fi)

tool holder

stålholder[nb], stålhaldar[nn] (nb) stálledoalan (se) stålhållare, stålfäste (sv) teränpidin (fi)

toolmaker

verktøymaker[nb], verktøymakar[nn] (nb) neavvorávdi (se) verktygsmakare (sv) työkaluntekijä (fi)

tool path

verktøybane (nb) neavvojohtolat (se) verktygsväg (sv)

tooth depth

tannhøyde[nb], tannhøgd[nn] (nb) bátneallodat (se) kugghöjd (sv) hampaan korkeus (fi)

toothed bar

See "rack, pitch rack, toothed bar".

toothed wheel

See "gear wheel, cogwheel, toothed wheel".

tooth pitch

See "circular pitch, tooth pitch".

torque, moment

moment (nb) momeanta (se) moment (sv) momentti (fi)

torque, twisting moment, turning moment

dreiemoment, vrimoment (nb) jorranmomeanta, botnjanmomeanta (se) vridmoment (sv) vääntömomentti (fi)

torque wrench

momentnøkkel (nb) momeantačoavdda (se) momentnyckel (sv) momenttiavain (fi)

toughness

seighet[nb], seigleik[nn] (nb) dávggasvuohta, sitkatvuohta (se) seghet (sv) sitkeys (fi)

track

See "belt, track".

track roller bearing, rigid ball bearing

sporkulelager (nb) gurraluođđaláger (se) spårkullager (sv) urakuulalaakeri (fi)

tractor

traktor (nb) traktor (se) traktor (sv) traktori (fi)

transformer

transformator (nb) transformáhtor (se) transformator (sv) muuntaja (fi)

transistor

transistor (nb) transistor (se) transistor (sv) transistori (fi)

transition fit

mellompasning (nb) gaskaheivehus (se) mellanpassning (sv) välitiukkuus (fi)

transmission

See "gearbox, transmission".

transmission

overføring, transmisjon (nb) sirdin, transmišuvdna (se) transmission, överföring (sv) siirto, siirtymä (fi)

transmission belt, driving belt

drivreim (nb) jođihanruopma (se) drivrem (sv) tehonsiirtohihna, käyttöhihna (fi)

transom

See "traverse, transom, cross beam, cross bar".

transporter

See "conveyor, transporter".

trapezoid thread

trapesgjenge (nb) trapesajeaŋga (se) trapetsgänga (sv) trapetsikierre (fi)

trash barrel

See "garbage can, trash barrel, dustbin".

travelling trolley, crane carriage

traversvogn, løpekatt (nb) doaresbielkávávdna (se) traversvagn (sv) nostovaunu, juoksuvaunu (fi)

traverse crane, overhead travelling crane

traverskran, løpekran (nb) doaresbielkálovtton (se) traverskran (sv) siltanosturi (fi)

traverse, transom, cross beam, cross bar

travers (nb) doaresbielká (se) travers, tvärbalk (sv) silta, poikkiparru, poikittaispalkki (fi)

treat

See “prepare, process, treat”.

triangular file

trekantfil (nb) golmmačiegat fiilu, golmmaborat fiilu (se) trekantsfil (sv) kolmikulmaviila (fi)

trim

trimme (nb) hárjehahttit (se) trimma (sv) virittää (fi)

trimming machine

See “folding machine, trimming machine”.

trunnion

See “gudgeon, pivot, journal, trunnion”.

try square

See “square, try square, joint hook”.

turbine

turbin (nb) turbiidna (se) turbin (sv) turbiini (fi)

turbo charger, turbo compressor

turbo, turbokompressor (nb) turbo (se) turbo (sv) turbo, ahdin (fi)

turbo compressor

See “turbo charger, turbo compressor”.

turbulence

turbulens (nb) jorri (se) turbulens, virvelbildning (sv) turbulensi, pyörteisyys (fi)

turn

dreie (nb) várvet, jorahit, vátnat (se) svarva (sv) sorvata (fi)

turner

dreier[nb], dreiar[nn] (nb) várvvár (se) svarvare (sv) sorvari, sorvaaja (fi)

turning moment

See “torque, twisting moment, turning moment”.

turnings, shavings

flytespon (nb) joatkkavuolahas (se) långspån (sv)

turning tool

See “lathe tool, turning tool”.

turn (of a winding)

vinding (nb) giesastat (se) lindningsvarv (sv) käämi, johdinkierros (fi)

tweezers

See “pinzers, tweezers, forceps”.

twist

See “wind, coil, spool, twist”.

twist drill, spiral drill

spiralbor (nb) bonjisbovra (se) spiralborr (sv) kierukkapora (fi)

twisting moment

See "torque, twisting moment, turning moment".

two component glue

herdelim, tokomponentlim (nb) buoššodanliibma, guovtteoasát liibma (se) härdlim, tvåkomponentlim (sv) kaksikomponenttiliima (fi)

two-stroke engine

totaktsmotor (nb) guovttedávttatmohtor (se) tvåtaktsmotor (sv) kaksitahtimoottori (fi)

tyre

dekk (nb) deahkka (se) däck (sv) rengas (fi)

unbalance

ubalanse (nb) dássehisvuohta (se) obalans (sv) epätasapaino (fi)

union

union, kopling (nb) lihtolaš, uniuvdna (se) koppling (sv) liitin (fi)

unit of measure

målenhet[nb], måleining[nn] (nb) mihttoovttadat (se) måttenhet (sv) mittayksikkö (fi)

universal joint

universalledd (nb) lášmeslađas (se) universalknut, kardanknut (sv) yleisnivel (fi)

universal milling machine

universalfresemaskin (nb) oppalašjursanmašiidna (se) universalfräsmaskin (sv) yleisjyrsinkone (fi)

universal setting gauge

kombinasjonsvinkel (nb) lotnolasviŋkil (se) yleiskulmamittain (fi)

up milling

See "conventional milling, up milling".

upright drilling machine, vertical drilling machine, pillar drilling machine

søyleboremaskin (nb) čuoldabovrenmašiidna (se) pelarborrmaskin (sv) pylväsporakone (fi)

upset

See "close a rivet, upset, clinch, jalt".

vacuum cleaner, dust cleaner

støvsuger[nb], støvsugar[nn] (nb) noavki, gavjanjaman (se) dammsugare (sv) pölynimuri (fi)

valve

ventil (nb) ventiila (se) ventil (sv) venttiili (fi)

valve disc

See "flap, damper, valve disc".

valve lever

See "valve rocker, valve lever".

valve rocker, valve lever

vippearm (nb) sugadangiehta (se) vipparm (sv) venttiilinvipu (fi)

valve seat

ventilsete (nb) ventiilačohkkehat (se) ventilsäte (sv) venttiilin istukka (fi)

valve tappet

ventilløfter[nb], ventilløftar[nn] (nb) ventiilalovttan (se) ventillyftare (sv) venttiilinnostin (fi)

vane

See "blade, scoop, bucket, vane".

variator

variator (nb) variáhtor (se) variator (sv) variaattori, muunnin (fi)

V-belt

See "conebelt, V-belt".

vehicle

kjøretøy (nb) vuodjinfievru (se) fordon (sv) ajoneuvo (fi)

velocity

See "speed, rate, velocity".

ventilation, venting, airing, aeration

ventilasjon (nb) áibmomolsun (se) ventilation (sv) ilmanvaihto (fi)

venting

See "ventilation, venting, airing, aeration".

vent pipe

See "flue, outlet, vent pipe, hood".

vernier

nonie, nonius (nb) nonie, noniusa (se) nonieskala, nonie (sv) noonio (fi)

vernier caliper

See "sliding caliper, vernier caliper, slide gauge".

vertical

See "perpendicular, vertical".

vertical drilling machine

See "upright drilling machine, vertical drilling machine, pillar drilling machine".

vibrate

vibrere (nb) sparaidit (se) vibrera (sv) täryttää, täristä (fi)

vibration

vibrasjon, sperring (nb) sparaideapmi (se) vibration (sv) tärinä (fi)

vice

See "bench vice, vice".

view

riss (nb) govadat (se) skiss (sv) luonnos (fi)

viscosity

viskositet (nb) viskositehta (se) viskositet (sv) viskositeetti (fi)

void

See "pore, void".

voltage

spenning (nb) gealdda (se) spänning (sv) jännite (fi)

wagon, carriage

vogn (nb) vávdna (se) vagn (sv) vaunu (fi)

washer

See "disc, plate, washer".

washer

stoppskive (nb) rovvi (se) bricka (sv) aluslaatta (fi)

washing machine

See "scouring machine, washing machine".

washing nozzle, rinsing nozzle

spylespiss (nb) cirggan-junni, cirggangeahči (se) munstycke (sv) suutin, suutinventtiili (fi)

waste rock

See "shale, waste rock".

water cooled

vann(av)kjølt[nb], vass(av)kjølt[nn (nb) čáhcečoaskuduvvon (se) vattenkyld (sv) vesijäähdytteinen (fi)

water level

See "level, water level".

water pump pliers

vannpumpetang[nb], vasspumpetong[nn] (nb) čáhcebumpedoaŋggat (se) vattenpumptång (sv) vesipumppupihdit, moniotepihdit (fi)

water separator

vannutskiller[nb], vatnutskiljar[nn] (nb) čáhcesirrehat (se) vattenavskiljare (sv) vedenerotin (fi)

wave length

bølgelengde[nb], bølgjelengd[nn] (nb) bárroguhkkodat (se) våglängd (sv) aallonpituus (fi)

wear

slitasje (nb) gollu, loaktu (se) slitage, nötning (sv) kuluminen (fi)

wearing steel

slitestål, styrejern (nb) oalásruovdi, loaktinstálli (se) slitstål (sv) olasrauta, ohjausrauta (fi)

wear resistance

slitestyrke (nb) loaktinnanosvuohta, loaktinnannodat (se) nötningsmotstånd, slitstyrka (sv) kulumiskestävyys (fi)

weave, oscillate

pendle (nb) dearbbadit, heailut (se) pendla (sv) heilua (fi)

web

See "rib, fin, web, gill".

wedge, machine key

kile, sporkile (nb) lohti (se) kil (sv) kiila (fi)

weld

sveis (nb) sveisa (se) svets (sv) hitsi (fi)

weld

sveise (nb) sveiset, áhcahit (se) svetsa (sv) hitsata (fi)

welder

sveiser[nb], sveisar[nn] (nb) sveisejeaddji (se) svetsare (sv) hitsaaja (fi)

welding

sveising (nb) sveisen (se) svetsning (sv) hitsaus (fi)

welding apparatus

See "welding machine, welding apparatus".

welding burner, welding torch

sveisebrenner[nb], sveisebrennar[nn] (nb) sveisenboalddan (se) svetsbrännare (sv) hitsauspoltin (fi)

welding electrode

sveiseelektrode (nb) sveisenelektroda (se) svetselektrod (sv) hitsauselektrodi (fi)

welding helmet

sveisemaske, sveisehjelm (nb) sveisenhápma, sveisenmáska (se) svetshjälm, svetsmask (sv) hitsauskypärä, hitsaussuojus (fi)

welding machine, welding apparatus

sveiseapparat (nb) sveisenapparáhta (se) svetsaggregat (sv) hitauslaite (fi)

welding seam

sveisefuge (nb) sveisenluodda, sveisengurra (se) svetsfog (sv) hitsisauma? (fi)

welding torch

See "welding burner, welding torch".

welding transformer

sveiseomformer[nb], -omformar[nn], -transformator (nb) sveisentransformáhtor, sveisenrievdadeaddji (se) svetsomformare (sv) hitsausmuuntaja (fi)

welding wire

sveisetråd (nb) sveisenárpu (se) svetstråd (sv) hitsauslanka (fi)

weld penetration

innsmelting (nb) sisasuddan, sisasuddadeapmi (se) inträngning (sv) tunkeuma (fi)

weld pool

See “molten pool, weld pool”.

wheel

hjul (nb) juvla, ráttis (se) hjul (sv) pyörä (fi)

wheel nut wrench

hjulkryss, kryss, felgkryss (nb) ruossa (se) fälgnyckel (sv) ristikkoavain (fi)

wheel puller

See “puller, wheel puller”.

whetstone, finishing stick

bryne, hein, brynestein (nb) sadjin, sadjingeađgi, dávžžan (se) bryne (sv) hiomakivi, kovasin (fi)

whetting

bryning, heining (nb) sadjin, dávžan (se) bryning, hening (sv) teroittaminen, hiominen (fi)

wick

veke[nb], veike (nb) veaika (se) veke (sv) sydän (fi)

wick lubrication

vekesmøring[nb], veikesmørjing[nn] (nb) veaikavuoidan (se) veksmörjning (sv) sydänvoitelu (fi)

width across flats, span of jaws

nøkkelvidde[nb], nøkkelvidd[nn] (nb) čoavddagalljodat (se) nyckelvidd, nyckelgap (sv) avainväli, kita (fi)

winch

vinsj, spill[nb], spel[nn] (nb) vinte, vinta (se) vinsch, spel (sv) vintturi, nosturi (fi)

wind, coil, spool, twist

vikle (nb) giessat (se) linda (sv) käämiä (fi)

winding, coiling

vikling (nb) giesaldat, giesahat, giessu (se) rullning, lindning (sv) käämitys (fi)

wing nut

vingemutter[nb], vengemutter[nn] (nb) soadjemuhtter (se) vingmutter (sv) siipimutteri (fi)

wire

See “cable, wire, cord”.

wire

vaier (nb) stállebáddi, stálletoavva (se) vajer, wire (sv) vaijeri (fi)

wire drawing

trådtrekking (nb) árpogeassin (se) tråddragning (sv) langanveto (fi)

wiring diagram, circuit diagram

koplingsskjema (nb) laktingovus, goallustingovus (se) kopplingsschema, kretsschema (sv) kytkentäkaavio, kaaviopiirros (fi)

wood saw, billett saw

buesag[nb], bogesag[nn], vedsag (nb) čielgesahá (se) bågsåg (sv) kaarisaha (fi)

wood-screw

treskrue (nb) muorraskruvva (se) träskruv (sv) puuruuvi (fi)

working environment

arbeidsmiljø (nb) bargobiras (se) arbetsmiljö (sv) työympäristö (fi)

working machine

arbeidsmaskin (nb) bargomašiidna (se) arbetsmaskin (sv) työkone (fi)

working margin, machining allowance

arbeidsmonn, bearbeidingsmonn, -monn (nb) bargomunni (se) bearbetningstillägg, arbetsmån (sv) työvara (fi)

working stroke

See "power stroke, working stroke".

workpiece

arbeidsstykke (nb) bargoávnnas, ávnnastat (se) arbetsstycke (sv) työkappale (fi)

works

See "manufacturing plant, works, company".

workshop

verksted[nb], verkstad[nn] (nb) barggahat, divohat, dagahat (se) verkstad (sv) korjaamo (fi)

worm

snekkeskrue (nb) bonjisskruvva (se) snäckskruv (sv) kierukka (fi)

worm gear, worm wheel

snekkehjul, snekkedrev, skruedrev (nb) bonjisjuvla (se) snäckhjul (sv) kierukkapyörä (fi)

worm wheel

See "worm gear, worm wheel".

wrench

See "spanner, wrench".

yield point

flytegrense (nb) máhccanrádji (se) flytgräns, övre sträckgräns (sv) ylempi myötöraja (fi)

zero point

nullpunkt (nb) nullačuokkis (se) nolläge (sv) nollakohta, nollapiste (fi)

zink-plating, galvanization

forsinking, galvanisering (nb) sinkadeapmi, sinken (se) förzinkning, galvanisering (sv) sinkitys (fi)

Index

- jyväinen[fi], 16
(for)ureining[nn], 64
(kierteen) nousu[fi], 74
(laakerin) holkki[fi], 10
(olje)tåkesmører[nb], 58
(polttoaineen) jäänestoaine[fi], 18
(strøm)aggregat[nb], 2
(trykk)lufthammer[nb], 3

A
aallonpituus[fi], 96
abrasive cloth, 1
abrasive grain, 1
abrasive particle, 1
absorb, 1
absorbera[sv], 1
absorbere[nb], 1
absorberet[se], 1
absorboida[fi], 1
accelerator (pedal), 1
acceptance limit, 1
accumulator, 1
accuracy, 65
acetylen[nb], 1
acetylen[sv], 1
acetylena[se], 1
acetylene, 1
acid-proof stainless steel, 1
ackumulator[sv], 1, 7
acute angle, 1
adapter, 1
adapter sleeve, 1
adapter[nb], 1
adapter[se], 1
adapter[sv], 1
adapteri[fi], 1
adaptor, 1
adhesive, 8
adjust, 2
adjustable packing ring cutter, 2
adjustable reamer, 2
adjustable spanner, 2
adjustable wrench, 2
adjusting, 2
adjusting screw, 2
adjustment, 2, 40
admixturing material, 34
aducering[sv], 89
aducerjärn[sv], 52
aduseren[se], 89
adusering[nb], 89
aduserjern[nb], 52
aduserruovdi[se], 52
adusointi[fi], 89
adusoitu valurauta[fi], 52
advance, 33
aeration, 95
agglutinant, 8
aggregáhta[se], 2
aggregat[nb], 2
aggregat[sv], 2
aggregate, 2
agregaatti[fi], 2
agrigaatti[fi], 2

áhcagahttit[se], 4
áhcagastinárpu[se], 33
áhcagastingarra[se], 73
áhcagastinlámpá[se], 10
áhcagastit[se], 40
áhcahit[se], 97
ahdin[fi], 93
ahjo[fi], 38
áibmočoaskuduvvon[se], 3
áibmofilttar[se], 3
áibmomolsun[se], 95
áibmosirren[se], 2
áibmoveažir[se], 3
aiddolašvuohta[se], 65
áigodat[se], 47
aine[fi], 53
aines[fi], 53
air bleeding, 2
air blowing tube, 2
air cooled, 3
air filter, 3
air hammer, 3
air venting, 2
airing, 95
ajoneuvo[fi], 95
akku[fi], 1, 7
akkulaturi[fi], 7
akkumulaattori[fi], 1
akkumuláhtor[se], 1
akkumulator[nb], 1
aksel[nb], 77
akseli[fi], 77
akselin uritus[fi], 82
akselitappi[fi], 77
akseltapp[nb], 77
aksiaalilaakeri[fi], 90
aksiaalinen[fi], 5
aksiála[se], 5
aksiallager[nb], 90
aksiálláger[se], 90
aksiell[nb], 5
áksil[se], 77
aksilculci[se], 77
aksildáhppa[se], 77
aksling[nb], 77
ákšu[se], 5
alasin[fi], 4
alasveisen[se], 15
albue[nb], 29
álggalaš[se], 7
álgguviđá olju[se], 22
álgguviđá ruovdi[se], 61
álgo-[se], 7, 66
álgoávnnas[se], 69
álgobiire[se], 66
álgodáhppa[se], 88
álgogealdda[se], 66
álgogealdu[se], 30
alignment, 3
alipaine[fi], 51
alkeis-[fi], 7
alkeisvaraus[fi], 30
alkio[fi], 7
alladeatta[se], 44
alladeattacirggon[se], 44
allen key, 44
allodatsárggon[se], 86
allowance, 90
alloy, 3
aloituskierretappi[fi], 88
áloravda[se], 8
alternate, 13
alternating current, 3
aluslaatta[fi], 96
álvi[se], 38
ambolt[nb], 4
amplifier, 3
analog[nb], 3
analog[sv], 3
analoga[se], 3
analogalaš[se], 3
analogic, 3
analoginen[fi], 3
analogous, 3
angle, 3, 7
angle bar, 4

angle gauge, 4, 83
angle iron, 4
angle of thread, 4
angle screwdriver, 4
angular grinder, 4
angulometer, 4
anker[nb], 72
ankkuri[fi], 72
anlegg[nb], 63, 86
anleggsmaskin[nb], 19
anläggning[sv], 63
anläggningsmaskin[sv], 19
anlöpa[sv], 4
anløpe[nb], 4
anløping[nb], 4
anlöpning[sv], 4
anneal, 4
annealing, 4
anrika[sv], 31
anrike[nb], 31
anriking[nb], 31
anrikning[sv], 31
ansats[nb], 78
ansats[sv], 78
ansatsfil[nb], 36
ansatsfil[sv], 36
ansatsnippel[nb], 44
ansatsvinkel[nb], 84
anslagsvinkel[nb], 84
anti-knock value, 4
anturi[fi], 24
anvil, 4
apparáhta[se], 4
apparat[nb], 4
apparat[sv], 4
apparatus, 4
arbeidsmaskin[nb], 99
arbeidsmiljø[nb], 99
arbeidsmonn[nb], 99
arbeidsslag[nb], 65
arbeidsstykke[nb], 99
arbeidstakt[nb], 65
arbetsmaskin[sv], 99
arbetsmiljö[sv], 99
arbetsmån[sv], 99
arbetsstycke[sv], 99
arbetstakt[sv], 65
arborrhuvud[sv], 9
arc eye, 4
árja[se], 30
armatur[nb], 35
armatur[sv], 35
armature, 35
armatuuri[fi], 35
armatuvra[se], 35
armering[nb], 5
armering[sv], 5
armeringsstål[nb], 70
armeringsstål[sv], 70
armouring, 5
árpogeassin[se], 99
artisan, 21
árvogássa[se], 57
asentaa kutistamalla[fi], 78
asentaa[fi], 47
asentaja[fi], 54
asento[fi], 79
asetus[fi], 40
asetuskulma[fi], 77
asetyleeni[fi], 1
assemble, 5
assembly tong, 62
asteikko[fi], 73
asynchronous, 5
asynchronous motor, 5
asynkron[nb], 5
asynkron[sv], 5
asynkroninen moottori[fi], 5
asynkroninen[fi], 5
asynkronmohtor[se], 5
asynkronmotor[nb], 5
asynkronmotor[sv], 5
asynkruvnnalaš[se], 5
atninčilgehus[se], 53
atomisering[sv], 5
atomization, 5

atomizing, 5
atta[se], 24
attán[se], 24
atulat[fi], 62
augebolt[nn], 32
aura[fi], 64
aurinkokenno[fi], 81
auto[fi], 12
automaatio[fi], 5
automaatti[fi], 5
automaattinen[fi], 5
automaattiteräs[fi], 38
automáhta[se], 5
automáhtalaš[se], 5
automáhtastálli[se], 38
automasjon[nb], 5
automašuvdna[se], 5
automat[nb], 5
automat[sv], 5
automatic, 5
automation, 5
automation[sv], 5
automatiseren[se], 5
automatisering[nb], 5
automatisering[sv], 5
automatisk[nb], 5
automatisk[sv], 5
automatization, 5
automaton, 5
automatstål[nb], 38
automatstål[sv], 38
avainväli[fi], 98
avarrusistukka[fi], 9
avarruspää[fi], 9
avbalansere[nb], 6
avbitartong[nn], 23
avbitartång[sv], 23
avbitertang[nb], 23
avbrytar-[nn], 19
avbryterkontakt[nb], 19
ávdnen[se], 35
ávdnet[se], 65
avdragar[nn], 67
avdragare[sv], 67
avdrager[nb], 67
avennin[fi], 69
avfallskärl[sv], 39
avgas[sv], 31
avgasmätare[sv], 31
avgasrör[sv], 31
avgass[nb], 31
avgassanalysator[nb], 31
avgassanlegg[nb], 31
avgassmålar[nn], 31
avgassmåler[nb], 31
avgassystem[sv], 31
avisoleringstang[nb], 86
avisoleringstong[nn], 86
ávjjoheapmi[se], 8
ávjočiehka[se], 22
ávju[se], 29
avlopp[sv], 59
avloppsrör[sv], 26
avlufting[nb], 2
avluftning[sv], 2
avløp[nb], 59
avløpsrør[nb], 26
ávnnas[se], 53
ávnnastat[se], 99
avolenkkiavain[fi], 36
avrettar[nn], 63
avretter[nb], 63
avrivare[sv], 63
avstickning[sv], 23
avstikking[nb], 23
avsug[nb], 37
avsugning[sv], 37
avsökare[sv], 73
avtappingsplugg[nb], 27
avtappningsplugg[sv], 27
avtrekk[nb], 37
avtrekker[nb], 67
avtrekkjar[nn], 67
avvik[nb], 25
avvikelse[sv], 25
awl, 5

axe, 5
axel[sv], 77
axeltapp[sv], 77
axial, 5
axial[sv], 5
axiallager[sv], 90
axiell[sv], 5
axle, 77

B

back clearance, 5
back edge of knife, 5
back[sv], 48
backfire, 48
backlash, 6
back-rake angle, 6
back-rake surface, 6
backskiva[sv], 47
backventil[sv], 13
backväxel[sv], 70
báddesahá[se], 6
bádjebasttat[se], 38
badjeldeatta[se], 60
bádjesveisen[se], 38
bádjeveažir[se], 79
bádji[se], 80
bagadas[se], 39
bagadasčuohpan[se], 2
bagaldat[se], 39
báhkadit[se], 43
báhkasgierdil[se], 43
báhkkagieđahallan[se], 43
báhkkagierdevaš[se], 43
báhtter[se], 7
báhttergealddán[se], 7
báisaboaltu[se], 32
báisan[se], 31
báisaskuohppu[se], 32
bajásdoalahus[se], 52
bajildaslohkki[se], 23
bakaksel[nb], 69
bakaxel[sv], 69
bakdokke[nb], 88
bakhjul[nb], 69
bakhjul[sv], 69
bakke[nb], 48
bakkskive[nb], 47
bakslag[nb], 48
bakslag[sv], 48
baksläppningsvinkel[sv], 5
balance, 6
balans[sv], 6
balanse[nb], 6
balk, 7
balk[sv], 7
ball, 6
ball bearing, 6
ball cage, 6
ball hammer, 6
ball retainer, 6
balpress[sv], 43
banaanipistoke[fi], 6
banana plug, 6
bananbunci[se], 6
banankontakt[sv], 6
bananplugg[nb], 6
bananstikker[nb], 6
band, 27
band saw, 6
bandmål[nb], 54
bandmått[sv], 54
bandsag[nb], 6
bandsåg[sv], 6
bankefasthet[nb], 4
bankefastleik[nn], 4
bar, 42, 46
barggahat[se], 99
bargoávnnas[se], 99
bargobiras[se], 99
bargobumbá[se], 91
bargodákta[se], 65
bargomašiidna[se], 99
bargomunni[se], 99
bargoneavvu[se], 91
bárra[se], 46
barre[nb], 46

barrel, 24, 28
bárroguhkkodat[se], 96
base, 38
basic, 7
basismål[nb], 56
basmått[sv], 56
basttat[se], 91
basttoheapmi[se], 8
bátneallodat[se], 91
bátnejuohku[se], 15
bátnejuvla[se], 40
bátnelohku[se], 58
bátneráhtis[se], 40
bátnestággu[se], 68
batteri[nb], 7
batteri[sv], 7
batteriija[se], 7
batteriladar[nn], 7
batteriladdare[sv], 7
batterilader[nb], 7
battery, 7
battery charger, 7
baufil[nb], 42
bávkkehus[se], 32
bávkkiheapmi[se], 32
bázahasbohcci[se], 31
bázahasbohttu[se], 78
bázahasgarra[se], 79
bázahasgássa[se], 31
bázahasmihtádas[se], 31
bázahasrusttet[se], 31
bázahasveažir[se], 79
bead, 7
beaivvášsealla[se], 81
beaktu[se], 29
bealledagahat[se], 42
beallejođadas[se], 76
beallejorbafiilu[se], 42
bealljebunci[se], 28
bealljegohput[se], 28
bealljesuojan[se], 28
beam, 7
bearbeide[nb], 65
bearbeiding[nb], 35
bearbeidingsmonn[nb], 99
bearbeta[sv], 52, 65
bearbetning[sv], 35
bearbetningstillägg[sv], 99
bearing, 7
bearing bush, 7
bearing housing, 7
bearing shell, 7
beassansátni[se], 60
become blunt, 7
bed, 7
bedrift[nb], 53
beetle, 53
behaldar[nn], 19
beholder[nb], 19
behållare[sv], 19
belasta[sv], 51
belaste[nb], 51
belastning[nb], 51
belastning[sv], 51
belegg[nb], 16
belg[nb], 7
beliggenhet[nb], 79
belle[se], 77
bellečeahppi[se], 90
bellerávdi[se], 90
belleskárrit[se], 90
bellows, 7
belt, 7
belt drive, 7
belt pulley, 67
belte[nb], 7
beltemotorsykkel[nb], 80
belysning[nb], 45
belysningsstyrka[sv], 45
beläggning[sv], 16
bench vice, 7
bend, 7
bend[nb], 7
bensiidna[se], 39
bensiini[fi], 39
bensiinimoottori[fi], 61

bensin[nb], 39
bensin[sv], 39
bensinmohtor[se], 61
bensinmotor[nb], 61
bensinmotor[sv], 61
bent screwdriver, 4
benzine, 39
bereevne[nn], 12
beslag[nb], 35
beslag[sv], 35
betong[nb], 18
betong[sv], 18
betoni[fi], 18
betoniteräs[fi], 70
betoŋga[se], 18
bevegelse[nb], 56
bevel, 8
bevel gear, 8
bevel protractor, 4, 83
bevel rule, 8
bibliotek[sv], 26
biegg[nb], 75
biegg[sv], 75
bielká[se], 7
bihttá[se], 8
biila[se], 12
bil[nb], 12
bil[sv], 12
bildeleser[nb], 73
bildskärm[sv], 74
biletlesar[nn], 73
billett saw, 99
bin, 19
binaarinen[fi], 8
binára[se], 8
binary, 8
bindemedel[sv], 8
bindemiddel[nb], 8
binding agent, 8
binær[nb], 8
binär[sv], 8
biprodukt[nb], 11
biprodukt[sv], 11
birrajohtu[se], 23
biskäregg[sv], 75
bistevaš[se], 61
bit, 8
bit[nb], 8
bit[sv], 8
bitti[fi], 8
bjelke[nb], 7
blacksmith, 80
blad[sv], 8
blade, 8
bladfjäder[sv], 50
bladfjær[nb], 50
bladfjør[nn], 50
bladmått[sv], 33
bladsøker[nb], 33
blandare[sv], 55
blandebatteri[nb], 55
blandekammer[nb], 55
blandingsforhold[nb], 67
blandningsförhållande[sv], 67
blandningskammare[sv], 55
blast furnace, 8
bleck[sv], 77
blecksax[sv], 90
bleckslagare[sv], 90
blekk[nn], 77
blekksaks[nn], 90
blekkslagar[nn], 90
bli oskarp[sv], 7
bli slö[sv], 7
blikk[nb], 77
blikkenslager[nb], 90
blikksaks[nb], 90
blindnagle[nb], 64
blindnit[sv], 64
blinker[sv], 47
blinkers[sv], 36
blinkljus[sv], 47
blinklys[nb], 36, 47
block, 8
block[sv], 8
blocktyg[sv], 88

blohkka[se], 8
blokk[nb], 8
blunt, 8
blunt file, 36
blyackumulator[sv], 49
blyakkumulator[nb], 49
blybatteri[nb], 49
blybatteri[sv], 49
blyfri bensin[nb], 49
blyfri bensin[sv], 49
blære[nb], 65
bløtlodding[nb], 81
blåsebelg[nb], 7
blåsespiss[nb], 2
boagán[se], 7
boaldámuš[se], 39
boaldin[se], 17
boaldinmohtor[se], 17
boallobeavdi[se], 48
boaltu[se], 9
boazzi[se], 10
boazzuhuhttit[se], 15
body, 9
bogesag[nn], 42, 99
bohccebargi[se], 64
bohccebasttat[se], 62
bohccečuohpan[se], 62
bohccejeaŋga[se], 62
bohccesággi[se], 64
bohcci[se], 62
bohkat[se], 27
bolt, 9
bolt cutter, 9
bolt[nb], 9
boltsaks[nb], 9
bomspår[sv], 82
bond, 28
bonding, 29
bonjisbovra[se], 94
bonjisgurra[se], 75
bonjisjuvla[se], 99
bonjisskruvva[se], 99
bonte[se], 37
bontelakta[se], 68
bontet[se], 37
bor[nb], 27
borahanáksil[se], 33
borahanbumpa[se], 33
borahanjuvla[se], 68
borahanskruvva[se], 33
boraheapmi[se], 33
borahit[se], 33
bore, 27
bore[nb], 27
boremaskin[nb], 27
borhylse[nb], 88
boring head, 9
borkile[nb], 29
borr[sv], 27
borra[sv], 27
borrat[se], 31
borrchuck[sv], 88
borrmaskin[sv], 27
borste[sv], 10
bortnis[se], 88
boson[se], 33
bosonruopma[se], 33
boss, 45
bossunnjunni[se], 2
botkenbasttat[se], 23
botkkon[se], 87
botnedáhppa[se], 64
botnjanmomeanta[se], 91
botnjat[se], 76
botntapp[nn], 64
bottoming tap, 64
bovra[se], 27
bovraskuohppu[se], 88
bovrenmašiidna[se], 27
bovret[se], 27
bow, 9
bow saw, 42
box, 9
box wrench, 9, 81
brake, 9
brake block, 10

brake disc, 9
brake drum, 9
brake fluid, 10
brake pad, 10
brake shoe, 10
brandslang[sv], 35
brandsläckare[sv], 35
brannslange[nb], 35
brannslokkingsapparat[nb], 35
brass, 10
braze, 10
brekkjern[nb], 22
brems[nb], 9
bremse[nb], 9
bremsekloss[nb], 10
bremseskive[nb], 9
bremsesko[nb], 10
bremsetrommel[nb], 9
bremsevæske[nb], 10
brennbar[nb], 17
brennstoff[nb], 39
brennverdi[nb], 43
brensel[nb], 39
bricka[sv], 96
brittleness, 10
broach, 69, 83
broms[sv], 9
bromsa[sv], 9
bromsback[sv], 10
bromskloss[sv], 10
bromsskiva[sv], 9
bromstrumma[sv], 9
bromsvätska[sv], 10
brons[sv], 10
bronsa[se], 10
bronša[se], 10
bronse[nb], 10
bronze, 10
brotflate[nn], 87
brotsch[sv], 69
brotscha[sv], 69
brotsj[nb], 69
brotsje[nb], 69
brotspon[nn], 87
brottyta[sv], 87
bruddflate[nb], 87
bruddspon[nb], 87
bruksanvisning[nb], 53
bruksanvisning[sv], 53
bruksrettleiing[nn], 53
brush, 10
bryna[sv], 44
bryne[nb], 44, 98
bryne[sv], 98
brynestein[nb], 98
bryning[nb], 98
bryning[sv], 98
brytar[nn], 87
brytare[sv], 87
brytarspets[sv], 19
bryter[nb], 87
bräckjärn[sv], 22
brännbar[sv], 17
bränsle[sv], 39
bråkjøling[nb], 68
bucket, 8
buckrake, 43
buesag[nb], 42, 99
buffer, 10
buhtadit[se], 17
buiku[se], 77
built-up edge, 10
bukke[nb], 37
bukkemaskin[nb], 37
bulb, 10
buller[sv], 57
bult[sv], 9
bultsax[sv], 9
bulvarat[se], 65
bumpa[se], 67
bumpedoaŋggat[se], 62
bumper, 10
bumpet[se], 67
buncejođas[se], 64
bunci[se], 60

bunntapp[nb], 64
buohtalaslakta[se], 60
buohtalatgoallus[se], 78
buolihahtti[se], 17
buollángarra[se], 79
buođđoventiila[se], 39
buoššodanliibma[se], 94
buoššodanomman[se], 42
buoššodeapmi[se], 43
buoššodit[se], 43
burgit[se], 26
burr, 10
bush(ing), 10
buškosahá[se], 89
bussning[sv], 10
but strap, 35
butt welding, 10
butt welding bend, 11
buváhat[se], 14
buvihanventiila[se], 90
buvkosahá[se], 89
buvku[se], 77
bygel[sv], 9
bygg[nb], 19
bygge[sv], 19
byggnad[sv], 19
byggnadsverk[sv], 63
by-pass, 11
bypass valve, 11
bypass[sv], 11
bypassventil[sv], 11
by-product, 11
bälg[sv], 7
bæreevne[nb], 12
bärförmåga[sv], 12
bølgelengde[nb], 96
bølgjelengd[nn], 96
børste[nb], 10
bøssing[nb], 10
bøyel[nb], 9
bøyle[nb], 9
bågfil[sv], 42
bågsåg[sv], 42, 99

C

cable, 11
cable clip, 11
cable shoe, 11
caggejuvla[se], 68
čáhcebumpedoaŋggat[se], 96
čáhcečalbmi[se], 50
čáhcečoaskuduvvon[se], 96
čáhcesirrehat[se], 96
čáhkanrádji[se], 1
cahkehahtti[se], 17
cahkkehanboksa[se], 13
cahkkehangiesttus[se], 45
cahkkehanginttal[se], 82
cahkkehanrusttet[se], 45
čálán[se], 66
čalbmeboaltu[se], 32
calibrate, 11
calipers, 26
calorific value, 43
camshaft, 11
čađagolganmihtádas[se], 36
čađđa[se], 11
cap, 11
capacitive, 11
capacitor, 18
capacity, 11
capillary action, 11
capstan, 69
car, 12
carbon, 11
carburettor, 12
cardan joint, 12
carpenter, 12, 48
carpenter's bench, 12
carriage, 96
carrier, 12
carrying capacity, 12
čárva[se], 17
čárvamihtádas[se], 17
čárvenbasttat[se], 40
čárvendoaŋggat[se], 40
čárvenruovdi[se], 15

čárvenskuohppu[se], 1
čárvet[se], 17
case-hardening, 12
časkinbovrenmašiidna[se], 42
časkindávgadat[se], 45
časkinguhkkodat[se], 50
časkinváiddon[se], 78
cast, 12
cast iron, 12
castellated nut, 12
caster, 72
casting, 12
castle nut, 12
catch, 51
čatnanávnnas[se], 8
čavgaheivehus[se], 90
čávgenbelle[se], 51
čavgenmomeanta[se], 86
čavgenoalul[se], 15
čávgenskuohppu[se], 88
čávgenspelle[se], 51
cavitation, 12
CDI-boks[nb], 13
CDI-boksa[se], 13
CDI-box, 13
CDI-laatikko[fi], 13
CDI-rasia[fi], 13
čeabetgalljideapmi[se], 35
ceaggut[se], 61
ceahkeheapmi[se], 85
cement, 40
centre bit, 13
centre drill, 13
centre punch, 13
centring gauge, 13
centrumborr[sv], 13
centrumvinkel[sv], 13
chain, 13
chain drive, 13
chain lock, 13
chain saw, 65
chain wheel, 13
chamfer, 8
change gear, 13
charge, 13
chaser, 13
chasing tool, 13
check valve, 13
chemical combination, 14
chemical compound, 14
chilling, 68
chip, 14
chip breaker, 14
chipping, 14
chipwood, 14
chisel, 14
choke, 14, 90
choke[nb], 14
chokespjäll[sv], 14
chrome coat, 14
chrome plate, 14
chuck, 14
chuck wrench, 14
chuck[nb], 14
chuck[sv], 14
chucknyckel[sv], 14
čiehka[se], 3, 83
čiehkamihtádas[se], 4
čiehkaruovdi[se], 4
čiehkasirddan[se], 76
čiehkaskruvvaruovdi[se], 4
čiehkašliipa[se], 4
čielgesahá[se], 99
cikcenbasttat[se], 23
cinder, 79
circlip pliers, 14
circuit breaker, 87
circuit diagram, 99
circular grinding machine, 15
circular pitch, 15
cirggangeahči[se], 96
cirgganjunni[se], 96
cirggon[se], 58
cirgoventiila[se], 46
cirgun[se], 46
cirkelsåg[sv], 15

cirkular saw, 15
cirkulär delning[sv], 15
clad welding, 15
clamp, 9, 15
clamp arbor, 15
clamp frame, 15
clamping bush, 1
clamping jaw, 15
clamping plate, 15
claw coupling, 15
clean, 64
clean off burrs, 15
clearance angle, 15
clearance surface, 15
climb milling, 16
clinch, 16
close a rivet, 16
closed circuit, 16
clutch, 16
čoahkkisvuohta[se], 24
coakcečoavdda[se], 68
coal, 11
coarse file, 16
coarse-grained, 16
čoaskasahá[se], 17
čoaskudannjalbi[se], 20
čoaskudanrusttetiskkan[se], 24
coat, 16
coating, 16
čoavddagalljodat[se], 98
čoavddavuolggahus[se], 84
čoavdit[se], 51
cock, 16
coefficient of linear thermal expansion, 16
cogwheel, 40
čohkačiehka[se], 1
čohkkedat[se], 75
čohkket[se], 5
čohkkosággi[se], 88
čohkolaš[se], 18, 88
čohkolašvárven[se], 88
coil, 16, 98
coil[nb], 45
coiling, 98
coke, 16
cold saw, 17
collar, 12, 78–80
combination pliers, 17
combination wrench, 17
combustible, 17
combustion, 17
combustion engine, 17
company, 53
compass saw, 89
compasses, 26
compensate, 17
compress, 17
compressed air, 17
compression, 17
compression gauge, 17
compression tester, 17
compressor, 18
computer, 18
concentrate, 31
concentration, 31
concrete, 18
condensate, 18
condensation, 18
condensed water remover, 18
condenser, 18
conducting material, 18
conductor, 18
cone, 88
conebelt, 18
coned, 18
conical, 18
coniform, 18
connecting rod, 62
construction, 19, 24
construction machine, 19
construction steel, 19
construction works, 63
consumption, 19
contact, 80

contact breaker, 19
contact face, 19
contact gap, 19
contact point, 19
contact set, 19
contactor, 19
container, 19
control, 19, 46, 85
control engineering, 19
control technique, 20
control unit, 20
conventional milling, 20
converter, 20
conveyor, 20
coolant, 20
cooling system fitter, 20
coordinate, 20
copper, 20
cord, 11
corner iron, 4
corner radius, 20
corrode, 73
corrosion, 20
cotter, 83
counter nut, 21
counterbalance, 6
countershaft, 21
countersink, 21
countersunk head, 21
counterweight, 6
couple, 21
coupling, 21
cover, 11
crafter, 21
craftsman, 21
crane, 21
crane carriage, 92
crank, 21
crank pin, 21
crankcase, 21
crankshaft, 21
crimp, 78
cross bar, 93
cross beam, 93
cross chisel, 22
cross edge, 49
cross slide, 22
cross tool slide, 22
cross-slot screwdriver, 22
crow bar, 22
crude oil, 22
čuggen[se], 23
čuggenstálli[se], 80
čugget[se], 22
culci[se], 41
čuohpastat[se], 22
čuohppanávjučiehka[se], 77
čuohppančikŋodat[se], 23
čuohppanleaktu[se], 23
čuohppanskearru[se], 23
čuokkissveisen[se], 83
čuokkissveisenapparáhta[se], 83
čuoldabovrenmašiidna[se], 94
čuonan[se], 82
čuonancahkkehat[se], 39
čuovgadávgi[se], 29
čuovgafierpmádat[se], 52
čuovgapeara[se], 10
cuozza[se], 54
current, 22
cursor, 22
cushion, 22
cushioning, 22
cut, 22
cut off, 22
cut threads, 89
cutting, 14
cutting angle, 22
cutting burner, 23
cutting depth, 23
cutting die, 23
cutting disc, 23
cutting edge, 29
cutting edge angle, 77
cutting off, 23

cutting pliers, 23
cutting speed, 23
cutting torch, 23
čuvgehus[se], 45
cycle, 23
cyclic, 61
cylinder, 23, 24
cylinder block, 23
cylinder cover, 23
cylinder head cover, 23
cylinder leak tester, 24
cylinder[sv], 23
cylinderblock[sv], 23
cylinderhuvud[sv], 23
cylindrical cutter, 24

D

dagahat[se], 99
dáhkkal[se], 88
dáhkohahtti[se], 38
dáhkut[se], 37
dáhppa[se], 41
dáhta[se], 24
dákta[se], 86
dammsugare[sv], 94
damp, 22
damp[nb], 85
damper, 36, 78, 79
dampmaskin[nb], 85
dárkilastin[se], 2
dárkilastit[se], 2
dárkivuohta[se], 65
dárkkistanbihttá[se], 40
dárkkistit[se], 19
dárkkistus[se], 46
dássádat[se], 6
dássehisvuohta[se], 94
dássenjuvla[se], 37
dásserávdnji[se], 25
dásset[se], 6
data, 24
data[fi], 24
data[nb], 24
data[sv], 24
datamaskin[nb], 18
datamaskin[sv], 18
datnejugaheapmi[se], 81
datni[se], 90
dator[sv], 18
dávgegealdda[se], 83
dávgerovvi[se], 73
dávgesahá[se], 42
dávgestálli[se], 83
dávggas[se], 29
dávggasbuoššodeapmi[se], 89
dávggasbuoššoduvvon stálli[se], 89
dávggasvuohta[se], 92
dávgi[se], 83
dávji[se], 38
dávžan[se], 98
dávžat[se], 44
dávžžan[se], 98
dead centre (lower / top), 24
dead centre peak, 24
dead centre point, 24
deaddeheivehus[se], 43
deaddemunni[se], 35
deaddinboallu[se], 67
deaddit[se], 65
deaeration, 2
deahkka[se], 94
deahpanit[se], 24
dearba[se], 60
dearbastihkka[se], 60
dearbbadit[se], 97
dearpannanosvuohta[se], 4
dearrebihttá[se], 46
deattaáibmu[se], 17
deattamihtádas[se], 66
deattán[se], 66
deattarádjenventiila[se], 66
deattareguláhtor[se], 66
deavddádas[se], 34
deavdinniibi[se], 82
deform, 24

deformation, 24
deformera[sv], 24
deformere[nb], 24
dekk[nb], 94
dekkgass[nb], 77
deksel[nb], 11
delehode[nb], 26
delehovud[nn], 26
delesirkel[nb], 62
deleskive[nb], 46
delevaskar[nn], 74
delevasker[nb], 74
deling[nb], 15
delningsdocka[sv], 26
delningshuvud[sv], 26
delningsskiva[sv], 46
demontera[sv], 26
demontere[nb], 26
dempe[nb], 22
demping[nb], 22
denna[se], 18
densimeter, 24
densitet[nb], 24
densitet[sv], 24
density, 24
depression, 51
design, 24
detaljlista[sv], 60
detector, 24
deviation, 25
device, 4
devkkodahttin[se], 4
devkkodahttit[se], 4
dial gauge, 25
dial indicator, 25
diaphragm, 54
die, 48, 57
die stock, 25
diesel engine, 25
diesel oil, 25
diesel[nb], 25
diesel[se], 25
dieselmohtor[se], 25
dieselmoottori[fi], 25
dieselmotor[nb], 25
dieselmotor[sv], 25
dieselolja[sv], 25
dieselolje[nb], 25
dieselolju[se], 25
dieselöljy[fi], 25
dievaslaš[se], 81
differensial[nb], 25
differensiála[se], 25
differensiálgiddehat[se], 25
differensialsperre[nb], 25
differential, 25
differential lock, 25
differential[sv], 25
differentialspärr[sv], 25
digitaalinen[fi], 25
digital, 25
digital[nb], 25
digital[sv], 25
digitála[se], 25
digitálalaš[se], 25
dihtor[se], 18
dimension, 25
dimensionera[sv], 25
dimensjonere[nb], 25
diod[sv], 25
dioda[se], 25
diode, 25
diode[nb], 25
diodi[fi], 25
direct current, 25
direction, 25
direction of current, 25
directional valve, 25
directory, 26
disc, 26
disc brake, 26
dismantle, 26
dismount, 26
display, 74
distributor, 26
divergence, 25

divider, 26
dividing head, 26
dividing plate, 46
divodeaddji[se], 54
divohat[se], 99
divvun[se], 70
divvut[se], 70
doaibma[se], 59
doaibmanmearri[se], 29
doaimmahis gássa[se], 46
doaimmár[se], 66
doalan[se], 44
doaŋggat[se], 91
doapparlihtti[se], 39
doaresávju[se], 49
doaresbielká[se], 93
doaresbielkálovtton[se], 92
doaresbielkávávdna[se], 92
doaresmielggas[se], 22
dobbeltverkande sylinder[nn], 26
dobbeltvirkende sylinder[nb], 26
doddjonolggoš[se], 87
doddjonvuolahas[se], 87
dollačuohpan[se], 23
dollaveažir[se], 19
domkraft[sv], 47
dor[nb], 27
dorn[sv], 27
double acting cylinder, 26
double row bearing, 28
dovddan[se], 24
dowel, 61
dowel saw, 89
down milling, 16
dragbrotsch[sv], 83
draghållfasthet[sv], 89
drain, 59
drain pipe, 26
drain plug, 27
drawing, 27
drawing board, 27
dredger, 31
dreiar[nn], 93
dreie[nb], 93
dreiebenk[nb], 49
dreiemoment[nb], 91
dreier[nb], 93
dreiestål[nb], 49
dreneringsplugg[nb], 27
dress, 31, 32, 64
dressed ore, 27
dresser, 63
dressing, 31
drev[nb], 28
drev[sv], 28
-dreven[nn], 39
drift, 27
drift[nb], 59
drift[sv], 59
drill, 27
drill coupling, 88
drilling head, 9
drilling machine, 27
drive, 27
drive fit, 90
drivhjul[nb], 28
drivhjul[sv], 28
driving belt, 27, 92
driving gear, 28
driving pinion, 28
driving wheel, 28
drivkraft[nb], 56
drivkraft[sv], 56
drivpasning[nb], 90
drivpassning[sv], 90
drivreim[nb], 92
drivrem[sv], 92
drivstoff[nb], 39
drivverk[nb], 27
drivverk[sv], 27
drum, 28
drum brake, 28
dual bearing, 28
dubb[sv], 88

dubbdocka[sv], 88
dubbelverkande cylinder[sv], 26
dubbhöjd[sv], 44
dubbrör[sv], 88
ductile, 28, 38
duhppen[se], 78
duhppet[se], 78
dulbenjurssán[se], 32
dulbenšliipenmašiidna[se], 86
dulbet[se], 63
dulka[se], 39
dull, 8
dulpensveisen[se], 10
dulpet[se], 16
dumájeaŋga[se], 46
dumámihtádas[se], 46
duobussillen[se], 34
duobussillet[se], 60
duobussilli[se], 34
duohpahuvvat[se], 78
duohpašuvvat[se], 78
duohppa[se], 33
duojár[se], 21
duolbabasttat[se], 36
duolbabeavdi[se], 87
duolbadávgi[se], 50
duolbajeaŋga[se], 36
duolbaluovččan[se], 36
duolbaruopma[se], 36
duolbaskearru[se], 32
duolbastálli[se], 36
duolbbas[se], 36
duolbbasinjorahit[se], 32
duolmman[se], 60
duolvačáhcebohcci[se], 26
duorrannávli[se], 71
duorranveažir[se], 6, 85
duorrat[se], 71
duŋke[se], 47
durra[se], 27
dust cleaner, 94
dustbin, 39
dustehat[se], 10
dynaaminen[fi], 28
dynámalaš[se], 28
dynamic, 28
dynamical, 28
dynamisk[nb], 28
dynamisk[sv], 28
dynamo, 40
dynamo[fi], 40
dynamo[nb], 40
dynamo[sv], 40
dysa[sv], 58
dyse[nb], 58
däck[sv], 94
dämpa[sv], 22
dämpning[sv], 22
dødgang[nb], 6
dödgång[sv], 6
dødpunkt (nedre / øvre)[nb], 24
dödpunkt (undre / övre)[sv], 24

E

e. shutdown, 30
eahpeguovddášlaš[se], 31
ear plug, 28
ear protection, 28
earmuff, 28
earth, 28
earth conductor, 29
earth lead, 29
earth wire, 29
earthing, 29
eccentricity, 29
edelgass[nb], 57
edge, 29
edge angle, 22
edge raising, 35
ednen[se], 29
ednenjođas[se], 29
ednet[se], 28
effect, 29

effekt[nb], 29
effekt[sv], 29
efficiency, 29
egg[nb], 29
egg[sv], 29
eggvinkel[nb], 22
eggvinkel[sv], 22
ejector, 29
eksentrisitet[nb], 29
eksentrisk[nb], 31
eksos[nb], 31
eksosa[se], 31
eksosanlegg[nb], 31
eksosbohcci[se], 31
eksosbohttu[se], 78
eksosgássa[se], 31
eksosmihtádas[se], 31
eksosmålar[nn], 31
eksosmåler[nb], 31
eksospotte[nb], 78
eksosrør[nb], 31
ekspansjon[nb], 31
ekspansjonsbolt[nb], 32
ekspansjonshylse[nb], 32
eksplosjon[nb], 32
elastic, 29
elastisk[nb], 29
elastisk[sv], 29
elbow, 7, 29
elbow bend, 29
el-čuggestat[se], 80
el-culci[se], 64
electric arc, 29
electric circuit, 30
electric(al), 29
electrician, 30
electricity, 30
electrode, 30
electronic, 30
electronics, 30
elektralaš[se], 29
elektricitet[sv], 30
elektrihkalaš[se], 29
elektrihkkár[se], 30
elektrikar[nn], 30
elektriker[nb], 30
elektriker[sv], 30
elektrisitehta[se], 30
elektrisitet[nb], 30
elektrisk aggregat[nb], 2
elektrisk[nb], 29
elektrisk[sv], 29
elektrod[sv], 30
elektroda[se], 30
elektrode[nb], 30
elektrodi[fi], 30
elektronihkka[se], 30
elektroniikka[fi], 30
elektronik[sv], 30
elektronikk[nb], 30
elektroninen[fi], 30
elektronisk[nb], 30
elektronisk[sv], 30
elektrovnnalaš[se], 30
elementary charge, 30
elementær[nb], 7
elementär[sv], 7
elementärladdning[sv], 30
elementærladning[nb], 30
elnät[sv], 52
emergency stop device, 30
emery cloth, 1
emery paper, 30
emne[nb], 53
emulsio[fi], 30
emulsion, 30
emulsion[sv], 30
emulsjon[nb], 30
end face mill, 77
end mill, 30
endeklaringsvinkel[nb], 5
endeplanfres[nb], 77
end-play angle, 5
energi[nb], 30
energi[sv], 30
energia[fi], 30

energiija[se], 30
energy, 30
engagement, 30
engine, 56
engine anchorage, 30
engine block, 23
engine bracket, 30
engine heater, 31
engine lug, 30
enkeltverkande sylinder[nn], 79
enkeltvirkende sylinder[nb], 79
enkelverkande cylinder[sv], 79
enrich, 31
enrichment, 31
ensisijainen[fi], 66
ensiöpiiri[fi], 66
epäkeskinen[fi], 31
epäpuhtaus[fi], 64
epätahtimoottori[fi], 5
epätasapaino[fi], 94
eretsveisen[se], 50
eriste[fi], 47
eristin[fi], 47
eristys[fi], 47
eristysnauha[fi], 47
eristää[fi], 47
erreldat[se], 47
erren[se], 47
errenbáddi[se], 47
erret[se], 47
erttet[se], 71
esijännitys[fi], 66
esikierretappi[fi], 88
essesveising[nb], 38
etch, 31
etsa[sv], 31
etsata[fi], 31
etse[nb], 31
etuakseli[fi], 39
etuvaunu[fi], 37
etuvetoinen[fi], 39
excavator, 31
excentric, 31
excentrisk[sv], 31
exchange, 40
exhaust, 31
exhaust analyzer, 31
exhaust box, 78
exhaust pipe, 31
exhaust system, 31
expanderbult[sv], 32
expanderhylsa[sv], 32
expansion, 31
expansion bolt, 32
expansion bush, 32
expansion sleeve, 32
expansion[sv], 31
expansion-shell bolt, 32
expansionshylsa[sv], 32
explosion, 32
explosion[sv], 32
extension bar, 32
extension cord, 32
extension rod, 32
eye bolt, 32
eyebolt, 32

F

fabrihkka[se], 32
fabrik[sv], 32
fabrikk[nb], 32
face, 32, 63
face chuck, 32
face plate, 32
facing, 16
facing cutter, 32
factory, 32
fadnil[se], 29
fállu[se], 54
fals[nb], 37
fals[sv], 37, 68
falsa[sv], 37
false[nb], 37

falsmaskin[sv], 37
falsskjøt[nb], 68
falsskøyt[nn], 68
fan, 33
fan belt, 33
fanasmohtor[se], 59
fápmočuohpan[se], 9
fápmonađđa[se], 87
fápmosirdin[se], 65
fápmoválddahat[se], 65
fárppal[se], 28
fárppalgoazan[se], 28
fart[nb], 82
fartsmålar[nn], 82
fartsmåler[nb], 82
fas[nb], 8
fas[sv], 8, 61
fasa[sv], 8
fase[nb], 8, 61
faseforskuving[nn], 61
faseforskyving[nb], 61
fasförskjutning[sv], 61
fasförslitning[sv], 61
faskkon[se], 74
faskut[se], 74
fasslitasje[nb], 61
fast dubb[sv], 24
fast senterspiss[nb], 24
fasten, 33
fasthet[nb], 86
fasthet[sv], 86
fastleik[nn], 86
fastnøkkel[nb], 59
fatigue, 33
fatnannannodat[se], 89
faucet, 16, 80
fault finding, 33
fault location, 33
fealga[se], 71
feed, 33
feed pump, 33
feed rod, 33
feed worm, 33
feeler gauge, 33
feilsøking[nb], 33
feittpresse[nn], 41
felg[nb], 71
felgkryss[nb], 98
felloe, 71
felly, 71
felsökning[sv], 33
felt, 33
felt[nb], 33
fertilizer distributor, 33
feste[nb], 33
festeskrue[nb], 76
fettpresse[nb], 41
fettspruta[sv], 41
field, 33
fiellosahá[se], 48
fierbmegealdda[se], 52
fierdin[se], 49
fierra[se], 71
fievrran[se], 20
fiidnagieđahallan[se], 34
fiidnagortnat[se], 34
fiidnajeaŋga[se], 34
fiila[se], 34
fiilet[se], 34
fiilogazza[se], 34
fiilogušta[se], 34
fiilu[se], 34
fil[nb], 34
fil[sv], 34
fila[sv], 34
filament, 33
filborste[sv], 34
filbørste[nb], 34
file, 34
file brush, 34
file[nb], 34
filing vice, 34
filklo[nb], 34
filklove[sv], 34
fill, 68
filler material, 34

fillet weld, 34
filling knife, 82
fillings, 34
filt[nb], 33
filt[sv], 33
filter, 34, 60
filter[nb], 34
filter[sv], 34
filtering, 34
filtra[sv], 60
filtration, 34
filtrere[nb], 60
filtrering[nb], 34
filtrering[sv], 34
filttar[se], 34
fin, 71
finbearbeiding[nb], 34
fine thread, 34
fine-grained, 34
finfördelning[sv], 5
fingjenge[nb], 34
fingänga[sv], 34
finish, 34, 64
finishing, 35
finishing stick, 98
finishing tap, 64
finkorna[nb], 34
finkornig[sv], 34
fire extinguisher, 35
fire hose, 35
firehjuling[nb], 38
firehjulssykkel[nb], 38
firetaktsmotor[nb], 38
firkantgjenge[nb], 36
firkantskrue[nb], 83
fish joint, 35
fit, 35
fitment, 35
fitnodat[se], 53
fittings, 35
fix, 33
fixing screw, 76
fixture, 35
fjernkontroll[nb], 70
fjernstyring[nb], 70
fjäder[sv], 83
fjäderbricka[sv], 73
fjäderringstång[sv], 14
fjäderspänning[sv], 83
fjäderstål[sv], 83
fjädrande rörpinne[sv], 64
fjær[nb], 83
fjärrmanövrering[sv], 70
fjærskive[nb], 73
fjærspenning[nb], 83
fjærsplint[nb], 42
fjærstål[nb], 83
fjør[nn], 83
fjørskive[nn], 73
fjørspenning[nn], 83
fjørsplint[nn], 42
fjørstål[nn], 83
flacktång[sv], 36
flackvinkel[sv], 83
flange, 35, 37
flanging, 35
flank, 15
flank of thread, 78
flankevinkel[nb], 4
flankvinkel[sv], 4
flap, 36
flare nut spanner, 36
flare nut wrench, 36
flash light, 36
flat, 36
flat chisel, 36
flat file, 36
flat nose pliers, 36
flat pliers, 36
flat steel, 36
flat thread, 36
flat transmission belt, 36
flat[nb], 36
flatgänga[sv], 36
flatjern[nb], 36
flatmeisel[nb], 36

flatmejsel[sv], 36
flatreim[nb], 36
flatrem[sv], 36
flatstål[nb], 36
flattang[nb], 36
flattong[nn], 36
fleirkilesamband[nn], 82
flens[nb], 35
flerkileforbindelse[nb], 82
flexible, 29
float, 36
float ball, 36
flotasjon[nb], 36
flotation, 36
flotation[sv], 36
flottør[nb], 36
flottör[sv], 36
flow direction, 25
flowmeter, 36
flue, 37
fluidizer, 37
fluks[nb], 37
fluksa[se], 37
fluksasuhkodat[se], 37
flukstetthet[nb], 37
flukstettleik[nn], 37
flussmedel[sv], 37
flussmiddel[nb], 37
flute, 49, 76
fluted pin, 37
flux, 37
flux density, 37
flytegrense[nb], 99
flytespon[nb], 93
flytgräns[sv], 99
flywheel, 37
fläkt[sv], 33
fläktrem[sv], 33
fläns[sv], 35
flöde[sv], 37
flödesmätare[sv], 36
flödestäthet[sv], 37
foarbma[se], 56
fog[sv], 41
foga[sv], 41
fold, 37
folding machine, 37
folding metre rule, 46
forage harvester, 37
foraksel[nb], 39
forbrenning[nb], 17
forbrenningsmotor[nb], 17
forbruk[nb], 19
forbränning[sv], 17
forceps, 62
fordelar[nn], 26
fordeler[nb], 26
fordon[sv], 95
forecarriage, 37
foretak[nb], 53
forgassar[nn], 12
forgasser[nb], 12
forge, 37, 80
forge hearth, 38
forge tongs, 38
forgeable, 38
forger, 80
forge-welding, 38
forging hammer, 79
forgjengetapp[nb], 88
forgreiningsrør[nb], 53
forhaustar[nn], 37
forhjuls-[nb], 39
forhøster[nb], 37
fork lift truck, 38
forkromme[nb], 14
forlenger[nb], 32
forlengjar[nn], 32
form[nb], 56
form[sv], 56
formbar[nb], 28
formendring[nb], 24
formförändring[sv], 24
forming, 38
forsenka hode[nb], 21
forsenka hovud[nn], 21

forsenkar[nn], 21
forsenke[nb], 21
forsenker[nb], 21
forsenkingsbor[nb], 21
forsinking[nb], 100
forspenning[nb], 66
forsterkar[nn], 3
forsterke[nb], 86
forsterker[nb], 3
forsterking[nb], 86
forsterkning[nb], 86
forstilling[nb], 37
forstøving[nb], 5
forstøvning[nb], 5
forurensning[nb], 64
fotpassar[nn], 60
fotpasser[nb], 60
found, 12
foundation, 38
foundry, 38
four wheel bike, 38
four-stroke engine, 38
framaksel[nb], 39
framaxel[sv], 39
frame, 71, 84
framhjulsdrevet[nb], 39
framhjulsdriven[sv], 39
framvagn[sv], 37
frasveising[nb], 50
free cutting steel, 38
free machining steel, 38
frekvens[nb], 38
frekvens[sv], 38
frequency, 38
fres[nb], 55
frese[nb], 55
fresemaskin[nb], 55
friction, 39
friction coupling, 39
friction tape, 47
friflate[nb], 15
friksjon[nb], 39
friksjonskopling[nb], 39
friktion[sv], 39
friktionskoppling[sv], 39
frivinkel[nb], 15
front axle, 39
front wheel drive, 39
fräs[sv], 55
fräsa[sv], 55
fräsmaskin[sv], 55
fråsveising[nn], 50
fuel, 39
fuge[nb], 41
fundament[nb], 38
fundament[sv], 38
funnel, 39
fuoladeapmi[se], 52
fuolahas[se], 14
fuolahus[se], 52
fuođđarrájan[se], 37
fuse, 39
fyllmedel[sv], 34
fyllmiddel[nb], 34
fyrhjuling[sv], 38
fyrkantskruv[sv], 83
fyrtaktsmotor[sv], 38
fälg[sv], 71
fälgnyckel[sv], 98
fält[sv], 33
fälthack[sv], 37
färdigbehandling[sv], 34
fästa[sv], 33
fästskruv[sv], 76
følar[nn], 24
føler[nb], 24
förbigång[sv], 11
förbigångsventil[sv], 11
förbrukning[sv], 19
förbränningsmotor[sv], 17
fördelare[sv], 26
företag[sv], 53
føretak[nn], 53
förgasare[sv], 12
förgreningsrör[sv], 53
förkroma[sv], 14

förlängningssladd[sv], 32
förlängningsstång[sv], 32
förorening[sv], 64
förspänning[sv], 66
förstärka[sv], 86
förstärkare[sv], 3
förstärkning[sv], 86
försänka[sv], 21
försänkare[sv], 21
försänkt huvud[sv], 21
förtapp[sv], 88
förtunningsmedel[sv], 89
förzinkning[sv], 100

G
gacoring[nb], 75
gaffaltrukka[se], 38
gaffellyftvagn[sv], 42
gaffeltruck[nb], 38
gaffeltruck[sv], 38
gáiddusstivren[se], 70
gáidu[se], 25
gákkan[se], 22
galján[se], 69
galjidit[se], 69
galkan[se], 48
gallehit[se], 73
gállen[se], 43
gállet[se], 43
galljudanoaivi[se], 9
galmmihanmekanihkkár[se], 20
galvanisering[nb], 100
galvanisering[sv], 100
galvanization, 100
garahanbasttat[se], 86
garasvuohta[se], 43
garbage can, 39
gardnjil[se], 29
garniture, 35
garradat[se], 43
garrajugahit[se], 10
garrametálla[se], 43
garramuitu[se], 42
garraskearru[se], 42
garvinrádji[se], 58
gas igniter, 39
gaskaáksil[se], 21
gaskačiehka[se], 15
gaskadáhppa[se], 75
gaskaheivehus[se], 92
gaskaolggoš[se], 15
gaskatčuohppan[se], 23
gaskatčuohppat[se], 22
gasket, 39
gaskkus[se], 78
gasoline, 39
gasoline motor, 61
gaspedal[sv], 1
gass-[nb], 39
gássaduolmman[se], 1
gássor[se], 12
gasspedal[nb], 1
gaständare[sv], 39
gate valve, 39
gauge, 39, 40, 54
gauge block, 40
gavjadeapmi[se], 5
gavjanjaman[se], 94
gávvaruovdi[se], 25
gazzadoaŋggat[se], 57
gazzalakta[se], 15
geahčebihttá[se], 57
geahčedoalan[se], 88
geahčedulbenjurssan[se], 77
geahčegaskačiehka[se], 5
geahpametálla[se], 50
geahpidanventiila[se], 69
gealdda[se], 89, 96
gealddehahtti[se], 69
gealdit[se], 13
geallat[se], 64
gear, 40
gear wheel, 40
gearbox, 40
gearing, 40

gearru[se], 26
gearsi[se], 35
geasán[se], 67
geavahančilgehus[se], 53
geavja[se], 42
geavli[se], 9
gehtegat[se], 13
generaattori[fi], 40
generáhtor[se], 40
generator, 40
generator[nb], 40
generator[sv], 40
geson[se], 37
giddehat[se], 14
giddehatčoavdda[se], 14
giddejuvvon rávdnjebiire[se], 16
giddenáksil[se], 15
giddenmekanisma[se], 51
giddenskruvva[se], 76
giddet[se], 33
giebahuhttit[se], 81
gieddi[se], 33
giehpa[se], 81
giehtaanulaš[se], 53
gieđahallan[se], 35
gieđahallat[se], 65
gieralohkki[se], 23
gierdočoavdda[se], 81
gierdodahkki[se], 26
gierdomihtádas[se], 60
gierdosahá[se], 15
giesahat[se], 98
giesaldat[se], 98
giesastat[se], 93
giessat[se], 98
giessu[se], 98
giesttus[se], 16
gievrudat[se], 3
gievrudit[se], 86
gievrudus[se], 86
giira[se], 40
giirakássa[se], 40
gilgačiehka[se], 4
gilgacikcenbasttat[se], 78
gill, 71
ginttalčoavdda[se], 82
ginttalgahpir[se], 82
ginttaljođas[se], 64
gižaldat[se], 44
gir[nb], 40
gire[nb], 13
girkasse[nb], 40
gitta guovddášnjunni[se], 24
givare[sv], 24
giver[nb], 24
gjenge[nb], 89
gjengebakke[nb], 23
gjengeflanke[nb], 78
gjengelære[nb], 89
gjengesnitt[nb], 23
gjengestigning[nb], 74
gjengestål[nb], 13
gjengesøker[nb], 89
gjengesøkjar[nn], 89
gjengetapp[nb], 88
gjengevinkel[nb], 4
gjennomstrømmingsmåler[nb], 36
gjennomstrøymingsmålar[nn], 36
gjevar[nn], 24
gjuta[sv], 12
gjuteri[sv], 38
gjutgods[sv], 12
gjutjärn[sv], 12
gjødselspreder[nb], 33
gjødselspreiar[nn], 33
gli[nb], 79
glida[sv], 79
glidelager[nb], 80
glidemotstand[nb], 80
glidlager[sv], 80
glidmotstånd[sv], 80
glow, 4, 40
glue, 40

glø[nb], 40
glöda[sv], 40
gløde[nb], 4, 40
glødeskal[nn], 73
glødeskall[nb], 73
glødetråd[nb], 33
glödga[sv], 4
glödlampa[sv], 10
glödskal[sv], 73
glödtråd[sv], 33
gneiste[nn], 82
gneistetennar[nn], 39
gniding[nb], 72
gnidning[sv], 72
gnist[nb], 82
gnista[sv], 82
gnisttenner[nb], 39
goahca[se], 39
goahcalakta[se], 39
goahcanbircu[se], 10
goahcanfárppal[se], 9
goahcangáma[se], 10
goahcanlohti[se], 10
goahcannjalbi[se], 10
goahcanskearru[se], 9
goahcat[se], 9
goaivunmašiidna[se], 31
goallus[se], 21
goallusjođas[se], 32
goallustingovus[se], 99
goallustit[se], 21
goavdi[se], 9
goavkedulka[se], 33
goazan[se], 9
goban[se], 21
gohpat[se], 21
gohpeloaktu[se], 62
gohpeoaivi[se], 21
gohppočoavdda[se], 9
gohppu[se], 9
gokčangeardi[se], 16
gokčanlakta[se], 60
gollu[se], 19, 96
golmmaborat fiilu[se], 93
golmmačiegat fiilu[se], 93
golmmamuddutmohtor[se], 90
golve[se], 81
gomuhanávju[se], 46
gožu[se], 81
gožuhuhttit[se], 81
gordnesturrodat[se], 40
govadat[se], 96
govččas[se], 11
govddon[se], 36
govdudeapmi[se], 36
governing, 40
governor, 70
govvalohkki[se], 73
grab tongs, 40
grad[nb], 10
grad[sv], 10
grada[sv], 15
grádaviŋkil[se], 4
grade[nb], 15
gradtapp[sv], 64
graduated cylinder, 40
gradvinkel[nb], 4
grafskrivar[nn], 64
grafskriver[nb], 64
grain size, 40
grannsemd[nn], 65
gravemaskin[nb], 31
grease, 51
grease gun, 41
grease nipple, 41
grensemål[nb], 50
grensesnitt[nb], 47
grind, 41
grind stone, 41
grinding, 41
grinding disc, 41
grinding gauge, 41
grinding machine, 41
grinding pin, 41
grinding wheel, 41
grip, 42

gripetong[nn], 40
griptang[nb], 40
griptång[sv], 40
grisepikk[nb], 74
groove, 41, 49, 76, 80
gropförslitning[sv], 62
gropslitasje[nb], 62
ground, 28
grounding, 29
grovbearbeiding[nb], 72
grovbearbetning[sv], 72
grovfil[nb], 16
grovfil[sv], 16
grovkorna[nb], 16
grovkornig[sv], 16
gränsmått[sv], 50
gränssnitt[sv], 47
grävmaskin[sv], 31
gråberg[nb], 77
gråberg[sv], 77
gudgeon, 41
guhkidangorri[se], 16
guide bar, 41
guksi[se], 8
gullosuojan[se], 28
gummi[nb], 72
gummi[se], 72
gummi[sv], 72
guoddináhppi[se], 12
guoddinnákca[se], 12
guorbmádeapmi[se], 51
guorbmádit[se], 51
guorusnatgealdda[se], 58
guorusnatjohtin[se], 45
guorusrassehit[se], 68
guovddášallodat[se], 44
guovddášbovra[se], 13
guovddášnjunni[se], 88
guovddášviŋkil[se], 13
guovlu[se], 25
guovttedávttatmohtor[se], 94
guovttedoaimmat sylinddar[se], 26
guovtteoasát liibma[se], 94
guovtteráiddoláger[se], 28
guđaborat čoavdda[se], 44
guđaborat skruvva[se], 44
guđačiegat čoavdda[se], 44
guđačiegat skruvva[se], 44
guđju[se], 47
gurra[se], 80
gurradit[se], 41
gurraluođđaláger[se], 92
gurraskruvva[se], 80
gurrenculci[se], 27
gusta[se], 10
gänga[sv], 89
gängback[sv], 23
gängflank[sv], 78
gängkloppa[sv], 25
gängpressande skruv[sv], 76
gängsnitt[sv], 23
gängstigning[sv], 74
gängstål[sv], 13
gängtapp[sv], 88
gängtolk[sv], 89
gängvinkel[sv], 4
gödselspridare[sv], 33
gå-grense[nb], 1
gångjärn[sv], 44

H

haarukkatrukki[fi], 38
hábmehahtti[se], 28
hack saw, 42
haft, 42
hair pin spring, 42
haka-avain[fi], 44
hakemisto[fi], 26
hakenøkkel[nb], 44
haknyckel[sv], 44
haldar[nn], 44
half finished goods, 42
half-round file, 42
háloravda[se], 8
hálteventiila[se], 25

hálti[se], 25
halvfabrikat[nb], 42
halvfabrikat[sv], 42
halvledare[sv], 76
halvleder[nb], 76
halvleiar[nn], 76
halvrundfil[nb], 42
halvrundfil[sv], 42
hamara[fi], 5
hammar[nn], 42
hammare[sv], 42
hammasjako[fi], 15
hammasluku[fi], 58
hammaspyörä[fi], 40
hammastanko[fi], 68
hammer, 42
hammer drill, 42
hammer[nb], 42
hampaan korkeus[fi], 91
hana[fi], 16
hand pallet truck, 42
hand riveter, 71
hand vice, 34
hand wheel, 42
handle, 21, 42
handtag[sv], 42
handtak[nb], 42
handverkar[nn], 21
hanger, 71
hankaaminen[fi], 72
hantverkare[sv], 21
haŋkinváiddon[se], 78
hápmemuhttin[se], 24
haponkestävä teräs[fi], 1
hard disc, 42
hard metal, 43
harddisk[nb], 42
harden, 43
hardening, 43
hardening furnace, 42
hardhet[nb], 43
hardleik[nn], 43
hardlodde[nb], 10
hardmetall[nb], 43
hardness, 43
harittaa[fi], 76
harja[fi], 10
harjapäävasara[fi], 85
hárjehahttit[se], 93
harkko[fi], 8, 46
harkkorauta[fi], 61
harppi[fi], 26
harrow, 43
harv[nb], 43
harv[sv], 43
hárva[se], 43
harvester, 57
haspel[nb], 69
haspel[sv], 69
hastighet[nb], 82
hastighet[sv], 82
hastighetsmätare[sv], 82
hastighetsmåler[nb], 82
hátna[se], 16
hauraus[fi], 10
hay press, 43
hay rake, 43
head stock, 43
heahteorustahtti[se], 30
heailu[se], 60
heailut[se], 97
hearing protection, 28
heat, 43
heat resistant, 43
heat treatment, 43
heating value, 43
heatproof, 43
heavval[se], 63
heavvalasttit[se], 63
heavvalbeaŋka[se], 12
heavvalmašiidna[se], 63
heavy force fit, 43
heavy shrink fit, 43
hehkua[fi], 40
hehkulamppu[fi], 10
hehkulanka[fi], 33

hehkuttaa[fi], 4
hehkutushilse[fi], 73
height of centres, 44
heilahduksen vaimennin[fi], 78
heilua[fi], 97
heiluri[fi], 60
hein[nb], 98
heine[nb], 44
heining[nb], 98
heinähäntä[fi], 43
heisekran[nb], 21
heitto[fi], 29
heivehus[se], 35
heivenviŋkil[se], 8
hela[fi], 35
hellingsvinkel[nb], 68
hellittää[fi], 51
heloitus[fi], 35
hendel[nb], 42
hengsel[nb], 44
hengsle[nb], 44
hening[sv], 98
herde[nb], 43
herdelim[nb], 94
herdeomn[nn], 42
herdeovn[nb], 42
herding[nb], 43
hevarm[nb], 50
hevstang[nb], 50
hevstong[nn], 50
hexagon nipple, 44
hexagon wrench, 44
hexagonal headed bolt, 44
hiekkapuhallus[fi], 73
hiekkopaperi[fi], 30
hienojyväinen[fi], 34
hienokierre[fi], 34
hienotyöstö[fi], 34
hierto[fi], 49
high pressure, 44
high pressure cleaner, 44
high-speed steel, 44
hihna[fi], 27
hihnakäyttö[fi], 7
hihnapyörä[fi], 67
hiiletyskarkaisu[fi], 12
hiili[fi], 11
hiiliharja[fi], 10
hiiri[fi], 56
hiljentää[fi], 22
hilseyly[fi], 73
hinge, 44
hioa[fi], 41, 44
hiomakangas[fi], 1
hiomakara[fi], 41
hiomakivi[fi], 41, 98
hiomakone[fi], 41
hiomalaikka[fi], 41
hiomarae[fi], 1
hiominen[fi], 41, 98
hirret[se], 76
hitauslaite[fi], 97
hitsaaja[fi], 97
hitsata[fi], 97
hitsattu palko[fi], 7
hitsaus[fi], 97
hitsauselektrodi[fi], 97
hitsauskypärä[fi], 97
hitsauslanka[fi], 98
hitsausmuuntaja[fi], 97
hitsauspoltin[fi], 97
hitsaussokeus[fi], 4
hitsaussuojus[fi], 97
hitsi[fi], 97
hitsisauma?[fi], 97
hitsisula[fi], 56
hiukkanen[fi], 60
hiusputki-ilmiö[fi], 11
hjul[nb], 98
hjul[sv], 98
hjulkryss[nb], 98
hoarbma[se], 56
-hohde[fi], 55
hohtimet[fi], 57
hoiganheivehus[se], 79

hoiganmihtádas[se], 80
hoist pulley, 88
holder, 44
holder[nb], 44
holga[se], 84
holkki[fi], 80
hollow punch, 44
holpipe[nn], 44
holsirkel[nn], 62
holtong[nn], 67
hona[sv], 44
hone, 44
hone[nb], 44
hood, 37
hook spanner, 44
hoonata[fi], 44
horisontal, 44
horisontal milling machine, 45
horisontal[nb], 44
horisontalfresemaskin[nb], 45
horisontalfräsmaskin[sv], 45
horisontell[sv], 44
horn, 45
horn[nb], 45
hose, 45
hose clamp, 45
hose clip, 45
hovedegg[nb], 52
hovtang[nb], 57
hovtong[nn], 57
hovtång[sv], 57
hovudegg[nn], 52
hub, 45
huggpipa[sv], 44
huksehus[se], 19
huksejeaddji[se], 48
hullpipe[nb], 44
hullsirkel[nb], 62
hulltang[nb], 67
hunet[se], 44
huokoinen[fi], 65
huokonen[fi], 65
huokosten muodostuminen[fi], 65
huolto[fi], 52
huopa[fi], 33
hurtigstål[nb], 44
huullos[fi], 37
huvudegg[sv], 52
huvudskäregg[sv], 52
hydraulalaš[se], 45
hydraulic, 45
hydraulics, 45
hydraulihkka[se], 45
hydrauliikka[fi], 45
hydraulik[sv], 45
hydraulikk[nb], 45
hydraulinen[fi], 45
hydraulisk[nb], 45
hydraulisk[sv], 45
hylkykivi[fi], 77
hylsa[sv], 9
hylse[nb], 10
hylsnyckel[sv], 9
hylsy[fi], 9
hylsyavain[fi], 9
hyvel[sv], 63
hyvelbänk[sv], 12
hyvelmaskin[sv], 63
hyvittää[fi], 17
hyvla[sv], 63
hyötysuhde[fi], 29
härda[sv], 43
härdlim[sv], 94
härdning[sv], 43
härdugn[sv], 42
hätäpysäytin[fi], 30
hävstång[sv], 50
høgderissar[nn], 86
høgresveising[nn], 71
högtryck[sv], 44
högtrycksaggregat[sv], 44
høgtrykk[nn], 44
høgtrykkspylar[nn], 44
hörselskydd[sv], 28

hörselskyddskåpa[sv], 28
hörselskyddspropp[sv], 28
hørselsvern[nb], 28
hösvans[sv], 43
høvel[nb], 63
høvelbenk[nb], 12
høvelmaskin[nb], 63
høvle[nb], 63
høyballepresse[nb], 43
høyderisser[nb], 86
höylä[fi], 63
höyläkone[fi], 63
höyläpenkki[fi], 12
höylätä[fi], 63
høyresveising[nb], 71
höyry[fi], 85
höyrykone[fi], 85
høysvans[nb], 43
høytrykk[nb], 44
høytrykkspyler[nb], 44
hålcirkel[sv], 62
hållare[sv], 44
hålstans[sv], 44
håltång[sv], 67
håndtak[nb], 42
håndtverker[nb], 21
hårddisk[sv], 42
hårdhet[sv], 43
hårdlöda[sv], 10
hårdmetall[sv], 43
hårnål[nb], 42
hårrørs-[nb], 11
hårrörsverkan[sv], 11

I
idle running, 45
idling, 45
ieš-[se], 82
iešcaggi skruvva[se], 76
iešdoalli skruvva[se], 76
iešguovddášahtti[se], 75
iešjeŋgenskruvva[se], 76
ignition coil, 45
ignition plug, 82
ignition system, 45
ikke-gå-grense[nb], 58
ikkje-gå-grense[nn], 58
ilá -[se], 60
illuminance, 45
illuminans[nb], 45
illuminans[sv], 45
ilmaisin[fi], 24
ilmajäähdytteinen[fi], 3
ilmanpoisto[fi], 2
ilmansuodatin[fi], 3
ilmanvaihto[fi], 95
impact resistance, 45
impact strength, 45
imperial thread, 46
impuls[nb], 46
impuls[sv], 46
impulsa[se], 46
impulse, 46
impulssi[fi], 46
imu[fi], 86
imusarja[fi], 53
inaktiv gass[nb], 46
inch rule, 46
inch thread, 46
index plate, 46
indexable insert, 46
indicator, 46
induce, 46
inducera[sv], 46
induction, 46
inductive, 46
induksjon[nb], 46
indukšuvdna[se], 46
induktiivalaš[se], 46
induktiivinen[fi], 46
induktio[fi], 46
induktion[sv], 46
induktiv[nb], 46
induktiv[sv], 46
indusere[nb], 46
induseret[se], 46

indusoida[fi], 46
inert gas, 46, 57
inert gas[sv], 46
inert gass[nb], 46
inerttikaasu[fi], 46
inflammable, 17
ingot, 46
ingrepp[sv], 30
injection, 46
injector, 46
injeksjon[nb], 46
injektor[nb], 46
inklinometer[sv], 4
inlet, 86
inner race, 46
inner ring, 46
innerring[nb], 46
innerring[sv], 46
inngrep[nb], 30
innsmelting[nb], 98
innsprøyting[nb], 46
innstillingsvinkel[nb], 77
innsuging[nb], 86
inre lagerring[sv], 46
inspection, 46
insprutning[sv], 46
insprutningsventil[sv], 46
install, 47
installera[sv], 47
installere[nb], 47
installoida[fi], 47
insulate, 47
insulating tape, 47
insulation, 47
insulator, 47
interface, 47
interference, 47
interferens[nb], 47
interferens[sv], 47
interferensa[se], 47
interferenssi[fi], 47
intermediate shaft, 21
intermittent, 61
intermittent light, 47
interrupter, 87
interval, 47
intervall[nb], 47
intervall[sv], 47
inträngning[sv], 98
iron, 47
iron bar, 22
iron ore, 47
irrotin[fi], 29, 67
irtosärmä[fi], 10
iskan[se], 89
iskunpituus[fi], 50
iskunvaimennin[fi], 78
iskuporakone[fi], 42
iskusitkeys[fi], 45
isolasjon[nb], 47
isolasjonstape[nb], 47
isolate, 47
isolation, 47
isolation[sv], 47
isolator[nb], 47
isolator[sv], 47
isolera[sv], 47
isolerband[nb], 47
isolerband[sv], 47
isolere[nb], 47
istuin[fi], 75
istukka[fi], 14
istukka-avain[fi], 14
itna[se], 39
itsekeskittävä[fi], 75
itsekierteittävä ruuvi[fi], 76
itsepidättävä ruuvi[fi], 76

J

jack, 47
jáddadanapparáhta[se], 35
jáddadanšláŋŋa[se], 35
jáffut[se], 65
jakkara[fi], 70
jako[fi], 15
jakoavain[fi], 2

jakohihna[fi], 69
jakoketju[fi], 69
jakolevy[fi], 46
jakopää[fi], 26
jakopään hihna[fi], 69
jakopään ketju[fi], 69
jakso[fi], 23
jaksoittainen[fi], 61
jalkapoljin[fi], 60
jállogássa[se], 57
jalokaasu[fi], 57
jalt, 16
jalusta[fi], 84
jamvekt[nb], 6
jápmačuokkis[se], 24
jargŋa[se], 35
jarru[fi], 9
jarrukenkä[fi], 10
jarrulevy[fi], 9
jarruneste[fi], 10
jarrupala[fi], 10
jarrurumpu[fi], 9
jarruttaa[fi], 9
jatke[fi], 32
jatkojohto[fi], 32
jatkovarsi[fi], 32
jauhe[fi], 65
jaw, 48
jaw chuck, 47
jeahkka[se], 47
jeaŋga[se], 89
jeaŋgagilga[se], 78
jeaŋgagoargŋun[se], 74
jeaŋgamihtádas[se], 89
jekk[nb], 47
jekketralle[nb], 42
jeŋgenbáhkka[se], 23
jeŋgendáhppa[se], 88
jeŋgenstálli[se], 13
jeŋget[se], 89
jern[nb], 47
jernmalm[nb], 47
jet, 58
jietnadorve[se], 45
jietnaváiddon[se], 78, 79
jig, 40, 48
jigg[nb], 48
jigg[sv], 48
jiki[fi], 48
joatkka[se], 32
joatkkajođas[se], 32
joatkkán[se], 32
joatkkavuolahas[se], 93
johde[fi], 7, 18
johdinkierros[fi], 93
johteet[fi], 7
johtimen liitin[fi], 11
johtinláger[se], 80
johtinvuosti[se], 80
johtit[se], 79
johto[fi], 11
johtoruuvi[fi], 49
johtu[se], 56
joiner, 48
joiner's saw, 48
joining, 48
joint, 37
joint hook, 83, 84
jointing, 39
jođadas[se], 18
jođas[se], 11
jođasávnnas[se], 18
jođasgáma[se], 11
jođasgámabasttat[se], 86
jođihandoaimmahat[se], 27
jođihanfápmu[se], 56
jođihanjuvla[se], 28
jođihanruopma[se], 92
jođihanrusttet[se], 27
jođihanskruvva[se], 49
jorahanbeaŋka[se], 49
jorahanruovdi[se], 12
jorahanskearru[se], 49
jorahit[se], 93
jorbabasttat[se], 72
jorbadas[se], 6

jorbafiilu[se], 72
jorbagrafihttaruovdi[se], 57
jorbašliipenmašiidna[se], 15
jorbaveažir[se], 6
jorda[sv], 28
jorde[nb], 28
jording[nb], 29
jordingsledning[nb], 29
jordingsleidning[nn], 29
jordledare[sv], 29
jordledning[nb], 29
jordleidning[nn], 29
jordning[sv], 29
jorggihansadji (bajit / vuolit)[se], 24
jorran[se], 70
jorrangierdu[se], 62
jorranlohku[se], 72
jorranmihtádas[se], 71
jorranmomeanta[se], 91
jorrat[se], 72
jorreáksil[se], 82
jorredoalan[se], 43
jorregeavri[se], 72
jorreláger[se], 72
jorreruovdi[se], 25
jorri guovddášnjunni[se], 50
jorri[se], 72, 93
journal, 41
journal bearing, 68
jousenjännitys[fi], 83
jousi[fi], 83
jousialuslaatta[fi], 73
jousialuslevy[fi], 73
jousiteräs[fi], 83
joustava[fi], 29
joutokäynti[fi], 45
jugahahttin[se], 81
jugahandatni[se], 81
jugaheapmi[se], 81
jugahit[se], 81
junction plate, 35
juogan[se], 26
juohkinoaivi[se], 26
juohkinskearru[se], 46
juoksute[fi], 37
juoksuvaunu[fi], 92
juotin[fi], 81
juotostahna[fi], 37
juottaa[fi], 81
juottaminen[fi], 81
juottokolvi[fi], 81
juottotina[fi], 81
jursanmašiidna[se], 55
jursat[se], 55
jurssahit[se], 55
jurssan[se], 55
justera[sv], 2
justerbrotsch[sv], 2
justere[nb], 2
justering[nb], 2
justering[sv], 2
justeringsskrue[nb], 2
justerskruv[sv], 2
juvla[se], 98
juvlaguovddáš[se], 45
juvlanáhpi[se], 45
jyrsin[fi], 55
jyrsinkone[fi], 55
jyrsiä[fi], 55
jännite[fi], 96
jännitys[fi], 89
järjestelmä[fi], 88
järn[sv], 47
järnmalm[sv], 47
järnskrot[sv], 74
jätekivi[fi], 77
jäykkyys[fi], 85
jäyste[fi], 10
jäähdytin[fi], 68
jäähdyttimen koestuslaite[fi], 24
jäähdytysneste[fi], 20

K

kaapeli[fi], 11

kaapelikenkä[fi], 11
kaapelikenkäpihdit [fi], 86
kaapia[fi], 74
kaari[fi], 71
kaarisaha[fi], 42, 99
kaasuhitsauspilli[fi], 11
kaasunsytytin[fi], 39
kaasupoljin[fi], 1
kaasutin[fi], 12
kaavin[fi], 74
kaaviopiirros[fi], 99
kabel[nb], 11
kabel[sv], 11
kabelsko[nb], 11
kabelsko[sv], 11
kabelskotang[nb], 86
kabelskotong[nn], 86
kabelskotång[sv], 86
kahva[fi], 42
kaivinkone[fi], 31
kaksikomponenttiliima[fi], 94
kaksirivinen laakeri[fi], 28
kaksitahtimoottori[fi], 94
kaksitoiminen sylinteri[fi], 26
kaksoisnippa[fi], 44
kaldsag[nb], 17
kalibrera[sv], 11
kalibrere[nb], 11
kalibreret[se], 11
kalibroida[fi], 11
kallistuskulma[fi], 68
kallsåg[sv], 17
kaltevuusmittari[fi], 4
kalvia[fi], 69
kalvin[fi], 69
kalvo[fi], 54
kamaksel[nb], 11
kamaxel[sv], 11
kamaxelkedja[sv], 69
kamaxelrem[sv], 69
kammentappi[fi], 21
kampi[fi], 21
kampiakseli[fi], 21
kampikammio[fi], 21
kamstål[nb], 70
kansi[fi], 11
kantavuus[fi], 12
kapacitet[sv], 11, 12
kapacitiv[sv], 11
kapasiteetti[fi], 11
kapasitehta[se], 11
kapasitet[nb], 11
kapasitiivalaš[se], 11
kapasitiivinen[fi], 11
kapasitiv[nb], 11
kapillaari-ilmiö[fi], 11
kapillardoaibma[se], 11
kapillaritet[sv], 11
kapillarvirkning[nb], 11
kapslipskiva[sv], 23
kara[fi], 82
karapylkkä[fi], 43
karbon[nb], 11
karbon[se], 11
karburatorsprit[sv], 18
kardang[nb], 12
kardanknut[sv], 12, 94
kardaŋga[se], 12
karhi[fi], 43
karkaista[fi], 43
karkaisu[fi], 43
karkaisusammutus[fi], 68
karkaisu-uuni[fi], 42
karkearakeinen[fi], 16
karkeatyöstö[fi], 72
karkeaviila[fi], 16
karkeus[fi], 72
kaross[sv], 9
karosseri[nb], 9
karosseri[sv], 9
kartio[fi], 88
kartiohammaspyörä[fi], 8
kartiokas[fi], 18
kartiosorvaus[fi], 88
kartiotappi[fi], 88
kast[nb], 29

kast[sv], 29
kasvu[fi], 31
katalog[nb], 26
katalog[sv], 26
katkaisin[fi], 87
katkaista[fi], 22
katkaisu[fi], 23
katkaisulaikka[fi], 23
katkojan kärki[fi], 19
katkojan kärkiväli[fi], 19
katkolastu[fi], 87
kauko-ohjaus[fi], 70
kaulus[fi], 79
kavitaatio[fi], 12
kavitasjon[nb], 12
kavitašuvdna[se], 12
kavitation[sv], 12
kedja[sv], 13
kedjedrift[sv], 13
kedjehjul[sv], 13
kedjelås[sv], 13
kehys[fi], 84
kela[fi], 16, 69
kelkka[fi], 79
kemiallinen sidos[fi], 14
kemiijalaš ovttastus[se], 14
kemisk förening[sv], 14
kenttä[fi], 33
keskimmäinen kierretappi[fi], 75
keskiökulma[fi], 13
keskiökärki[fi], 88
keskiöpora[fi], 13
kesto-[fi], 61
kestävyys[fi], 86
ketju[fi], 13
ketjukäyttö[fi], 13
ketjulukko[fi], 13
ketjupyörä[fi], 13
ketjut[fi], 13
kevytmetalli[fi], 50
key bed, 48
key groove, 48
keyboard, 48
keyway, 48
keyway cutter, 48
kick, 48
kierre[fi], 89
kierrekampa[fi], 89
kierreleuka[fi], 23
kierremitta[fi], 89
kierrepakka[fi], 23
kierresorkka[fi], 25
kierretappi[fi], 88
kierreterä[fi], 13
kierreura[fi], 75
kierroslukumittari[fi], 71
kierteen kylki[fi], 78
kierteittää[fi], 89
kierteitystyökalu[fi], 13
kierto[fi], 70
kiertokanki[fi], 62
kierukka[fi], 99
kierukkapora[fi], 94
kierukkapyörä[fi], 99
kiila[fi], 97
kiilahihna[fi], 18
kiilaholkki[fi], 32
kiilapultti[fi], 32
kiilaura[fi], 48
kiilauranhöylä[fi], 83
kiilauranjyrsin[fi], 48
kiillottaa[fi], 64
kiinnitin[fi], 44
kiinnittää[fi], 33
kiinnityskara[fi], 15
kiinnitysleuka[fi], 48
kiinnitysruuvi[fi], 76
kiinnitystuurna[fi], 15
kiinteä keskiökärki[fi], 24
kiintoavain[fi], 59
kiintolenkkiavain[fi], 17
kiintolevy[fi], 42
kil[sv], 97
kile[nb], 97
kilereim[nb], 18

kilreim[nb], 18
kilrem[sv], 18
kilspor[nb], 48
kilsporfres[nb], 48
kilsporhøvel[nb], 83
kilspår[sv], 48
kilspårfräs[sv], 83
kilspårsfräs[sv], 48
kilspårshyvel[sv], 83
kilsveis[nb], 34
kipinä[fi], 82
kiristysholkki[fi], 32, 88
kiristysleuka[fi], 15
kiristysmomentti[fi], 86
kirjoitin[fi], 66
kirves[fi], 5
kirvesmies[fi], 12, 48
kita[fi], 98
kitka[fi], 39
kitkakytkin[fi], 39
kjede[nb], 13
kjededrift[nb], 13
kjedehjul[nb], 13
kjedelås[nb], 13
kjemisk forbindelse[nb], 14
kjemisk sambinding[nn], 14
kjerringkjeft[nb], 57
kjetting[nb], 13
kjoks[nb], 14
kjoksnøkkel[nb], 14
kjølesystemtestar[nn], 24
kjølesystemtester[nb], 24
kjølevæske[nb], 20
kjøretøy[nb], 95
kjørnar[nn], 13
kjørner[nb], 13
klangkontroll[sv], 81
klangprøve[nb], 81
klaringsflate[nb], 15
klaringsvinkel[nb], 15
klemhylse[nb], 1
klemjern[nb], 15
klinke[nb], 71
klinknagle[nb], 71
kloakkrør[nb], 26
klokopling[nb], 15
klokoppling[sv], 15
klubba[sv], 53
klubbe[nb], 53
klämhylsa[sv], 1
kløtsj[nb], 16
knackningsbeständighet[sv], 4
knee, 7
knekke[nb], 37
knekkemaskin[nb], 37
knife, 49
knife tool, 49
kniv[nb], 49
kniv[sv], 49
knivrygg[nb], 5
knivrygg[sv], 5
knivstål[nb], 49
knivstål[sv], 49
knock rating, 58
knurl, 49
knurled pin, 37
knärör[sv], 7
koe[fi], 89
kofot[sv], 22
kohdistin[fi], 22
kohtisuora[fi], 61
koje[fi], 4
koks[nb], 16
koks[sv], 16
koksa[se], 16
koksi[fi], 16
kol[nn], 11
kol[sv], 11
kolmikulmaviila[fi], 93
kolmivaihemoottori[fi], 90
kolv[sv], 62
kolvring[sv], 62
kolvslag[sv], 86
kolvstång[sv], 62
kombinasjonsnøkkel[nb], 17

kombinasjonstang[nb], 17
kombinasjonsvinkel[nb], 94
kombinationstång[sv], 17
kompensera[sv], 17
kompensere[nb], 17
kompensoida[fi], 17
kompresjon[nb], 17
kompresjonsmålar[nn], 17
kompresjonsmåler[nb], 17
kompression[sv], 17
kompressionsmätare[sv], 17
kompressor[nb], 18
kompressor[se], 18
kompressor[sv], 18
kompressori[fi], 18
kompresuvdna[se], 17
kompresuvdnamihtádas[se], 17
komprimera[sv], 17
komprimere[nb], 17
komprimeret[se], 17
kon[nb], 88
kona[sv], 88
kondeansa[se], 18
kondeansajávkkadas[se], 18
kondens[nb], 18
kondens[sv], 18
kondensaattori[fi], 18
kondensáhtor[se], 18
kondensator[nb], 18
kondensator[sv], 18
kondenseren[se], 18
kondensering[nb], 18
kondensering[sv], 18
kondensfjernar[nn], 18
kondensfjerner[nb], 18
kondensointi[fi], 18
kondenssi[fi], 18
kondreiing[nb], 88
koneellistaminen[fi], 54
koneistaa[fi], 52
koneistaminen[fi], 52
koneisto[fi], 54
konepeitto[fi], 11
konisk dreiing[nb], 88
konisk tannhjul[nb], 8
konisk[nb], 18
konisk[sv], 18
koniskt kugghjul[sv], 8
konpinne[nb], 88
konpinne[sv], 88
konstruksjon[nb], 24
konstruksjonsstål[nb], 19
konstrukšuvdna[se], 24
konstruktio[fi], 24
konstruktion[sv], 24
konstruktionsstål[sv], 19
konsvarvning[sv], 88
kontakt[nb], 80
kontakt[sv], 80
kontaktavstand[nb], 19
kontaktavstånd[sv], 19
kontaktor[nb], 19
kontaktor[se], 19
kontaktor[sv], 19
kontaktori[fi], 19
kontaktåpning[nb], 19
kontramutter[nb], 21
kontramutter[sv], 21
kontroll[nb], 46
kontroll[sv], 46
kontrollera[sv], 19
kontrollere[nb], 19
konverter[nb], 20
konverter[se], 20
konverter[sv], 20
konvertteri[fi], 20
koordinaatti[fi], 20
koordináhta[se], 20
koordinat[nb], 20
koordinat[sv], 20
koota[fi], 5
kopar[nn], 20
kople[nb], 21
kopling[nb], 16, 21, 94
koplingsskjema[nb], 99

kopp[nb], 9
koppar[sv], 20
kopper[nb], 20
koppla[sv], 21
koppling[sv], 16, 21, 94
kopplingsschema[sv], 99
koppnøkkel[nb], 9
koputuskoe[fi], 81
kori[fi], 9
korjaaja[fi], 54
korjaamo[fi], 99
korjata[fi], 70
korjaus[fi], 70
korkeapaine[fi], 44
korkeapainepesulaite[fi], 44
kornstorleik[nn], 40
kornstorlek[sv], 40
kornstørrelse[nb], 40
korroosio[fi], 20
korrosion[sv], 20
korrosjon[nb], 20
korrošuvdna[se], 20
kortslutning[nb], 78
kortslutning[sv], 78
kortspån[sv], 87
korvakupit[fi], 28
korvakuvut[fi], 28
korvata[fi], 17
korvatulppa[fi], 28
kosketus[fi], 30
kovajuottaa[fi], 10
kovalevy[fi], 42
kovametalli[fi], 43
kovasin[fi], 98
kovuus[fi], 43
kraftarm[nb], 87
kraftavbitar[nn], 9
kraftavbitare[sv], 9
kraftavbiter[nb], 9
kraftoverføring[nb], 65
kraftuttag[sv], 65
kraftuttak[nb], 65
kraftöverföring[sv], 65
krage[nb], 78
kraging[nb], 35
kragning[sv], 35
kran[nb], 16, 21
kran[sv], 16
kretsschema[sv], 99
kromata[fi], 14
krommadit[se], 14
krommet[se], 14
kronemutter[nb], 12
kronmutter[sv], 12
krumcirkel[sv], 60
krumpassar[nn], 60
krumpassare[sv], 60
krumpasser[nb], 60
kruunumutteri[fi], 12
krympa på[sv], 78
krympa[sv], 78
krympe på[nb], 78
krympe[nb], 78
krymping[nb], 78
krympning[sv], 78
kryss[nb], 98
kryssmeisel[nb], 22
kryssmejsel[sv], 22
krysskruvmejsel[sv], 22
kubein[nb], 22
kuggdelning[sv], 15
kugghjul[sv], 40
kugghöjd[sv], 91
kuggrem[sv], 69
kuggstång[sv], 68
kuggtal[sv], 58
kula[sv], 6
kuldemontør[nb], 20
kule[nb], 6
kulegrafittjern[nb], 57
kulehaldar[nn], 6
kulehammar[nn], 6
kulehammer[nb], 6
kuleholder[nb], 6
kulelager[nb], 6
kulhammare[sv], 6

kulhållare[sv], 6
kuljetin[fi], 20
kull[nb], 11
kullager[sv], 6
kulma[fi], 3
kulmahiomakone[fi], 4
kulmamittari[fi], 4
kulmaruuvitaltta[fi], 4
kulmateräs[fi], 4
kulmaviivain[fi], 84
kuluminen[fi], 96
kulumiskestävyys[fi], 96
kumi[fi], 72
kunnossapito[fi], 52
kuolokohta (ala-/ylä-)[fi], 24
kuona[fi], 79
kuonahakku[fi], 79
kuonavasara[fi], 79
kuorintapihdit[fi], 86
kuormitettavuus[fi], 12
kuormittaa[fi], 51
kuormitus[fi], 51
kupari[fi], 20
kurvritare[sv], 64
kurvskrivare[sv], 64
kutistua[fi], 78
kutistuma[fi], 78
kutistuminen[fi], 78
kuttdjupn[nn], 23
kuttdybde[nb], 23
kutteskive[nb], 23
kuula[fi], 6
kuulakehikko[fi], 6
kuulalaakeri[fi], 6
kuulanpidin[fi], 6
kuulapäävasara[fi], 6
kuulonsuojain[fi], 28
kuumentaa[fi], 4, 43
kuusioavain[fi], 44
kuusioruuvi[fi], 44
kuvanlukija[fi], 73
kylare[sv], 68
kylkikulma[fi], 4
kyllästää[fi], 73
kylmontör[sv], 20
kylmäkoneasentaja[fi], 20
kylmäkonekorjaaja[fi], 20
kylmäsaha[fi], 17
kylvätska[sv], 20
kynsikytkin[fi], 15
kytkentäkaavio[fi], 99
kytkeä[fi], 21
kytkin[fi], 16, 21, 87
kädensija[fi], 42
kälsvets[sv], 34
kärkikorkeus[fi], 44
kärkikulma[fi], 1
kärkipylkkä[fi], 88
käsi-[fi], 53
käsikäyttöinen[fi], 53
käsisaha[fi], 48
käsityöläinen[fi], 21
kätting[sv], 13
käynnistin[fi], 84
käynnistyshammaskehä[fi], 84
käynnistysmoottori[fi], 84
käynnistää[fi], 84
käyrä[fi], 7
käyttö[fi], 19, 59
käyttöhihna[fi], 92
käyttölaite[fi], 27
käyttömekanismi[fi], 27
käyttöohje[fi], 53
käyttöpyörä[fi], 28
käyttövoima[fi], 56
käämi[fi], 16, 93
käämitys[fi], 98
käämiä[fi], 98
kääntökulmain[fi], 8
körnare[sv], 13
körriktningsvisare[sv], 47

L

laahia[fi], 44
laajeneminen[fi], 31
laakeri[fi], 7

laakerikuori[fi], 7
laakerin sisäkehä[fi], 46
laakerin ulkokehä[fi], 59
laakeripesä[fi], 7
laatta[fi], 63
ladata[fi], 13
ladda[sv], 13
láddjenmašiidna[se], 57
laddningsbar[sv], 69
lade[nb], 13
ladjoakkumuláhtor[se], 49
lager[nb], 7
láger[se], 7
lager[sv], 7
lágerbeassi[se], 7
lagerbox[sv], 7
lágergárri[se], 7
lagerhus[nb], 7
lagerhus[sv], 7
lagerskål[nb], 7
lagerskål[sv], 7
lagra[sv], 73
lagre[nb], 73
láhkki[se], 13
laipoitus[fi], 35
laippa[fi], 35, 41
laita[fi], 71
laite[fi], 4
laiteperustus[fi], 38
lajuhis bensiidna[se], 49
lakta[se], 47
laktingovus[se], 99
laktit[se], 21
lamealla[se], 63
lamell[nb], 63
lamell[sv], 63
lamelli[fi], 63
lamp bulb, 10
langanveto[fi], 99
lannanlevittäjä[fi], 33
lađascoavdda[se], 87
lapping, 49
larvband[sv], 7
lasihanbuoššodeapmi[se], 12
laskostuskone[fi], 37
laskskjøt[nb], 35
laskskøyt[nn], 35
láskut[se], 44
lášmeslađas[se], 94
lássenbelle[se], 75
lássenrieggabasttat[se], 14
lássenriekkis[se], 51
lássenskearru[se], 51
lasta[fi], 82
lastu[fi], 14
lastuamisnopeus[fi], 23
lastuamissyvyys[fi], 23
lastuava työstö[fi], 14
lastulevy[fi], 14
lastunmurtaja[fi], 14
lastupinta[fi], 6
lateral edge, 49
lathe, 49
lathe bed, 7
lathe carrier disc, 49
lathe tool, 49
latnalaslakta[se], 60
lattahihna[fi], 36
lattakierre[fi], 36
lattapihdit[fi], 36
lattataltta[fi], 36
lattateräs[fi], 36
lattaviila[fi], 36
laturi[fi], 40
lauhde[fi], 18
lauhtuminen[fi], 18
lausegg[nb], 10
lávgadangierdu[se], 75
lávgadanriekkis[se], 75
lávgadas[se], 75
lávgehat[se], 19
lávgenruovdi[se], 15
lávži[se], 27
lavtrykk[nb], 51
lavtta[se], 16, 21
lávvamuhttin[se], 24

lávvolat bátnejuvla[se], 8
lávvolat[se], 18, 88
lávvolatvárven[se], 88
lead accumulator, 49
lead battery, 49
lead screw, 49
lead-free gasoline / petrol, 49
leading screw, 49
leadless, 49
leaf spring, 50
leahttu[se], 82
leak, 50
leakage, 50
leaktomihtádas[se], 82
leaktu[se], 82
ledare[sv], 18
ledarskruv[sv], 49
leddhandtak[nb], 87
leddnøkkel[nb], 87
leder[nb], 18
ledeskrue[nb], 49
ledhandtag[sv], 87
ledhylsnyckel[sv], 87
ledning[nb], 11
ledning[sv], 11
leftward welding, 50
legera[sv], 3
legere[nb], 3
legering[nb], 3
legering[sv], 3
lehtijousi[fi], 50
leiar[nn], 18
leidning[nn], 11
leieskrue[nn], 49
leikehat[se], 38
leiket[se], 12
leikkaaminen[fi], 23
leikkaus[fi], 22
leikkauspoltin[fi], 23
leikkuri[fi], 74
leikkuripihdit[fi], 23
leikkurit[fi], 23
leikkuu[fi], 23
leikkuunopeus[fi], 23
leikkuusyvyys[fi], 23
leikkuusärmä[fi], 29
leikonávnnas[se], 12
leikonruovdi[se], 12
lejeerata[fi], 3
leka[fi], 79
lekk[nb], 50
lekkasje[nb], 50
lengdeutvidingskoeffisient[nb], 16
length of stroke, 50
lenkki[fi], 9
lenkkiavain[fi], 81
lepping[nb], 49
letku[fi], 45
letkunkiristin[fi], 45
lettmetall[nb], 50
leuansuojukset[fi], 51
leukaistukka[fi], 47
level, 44, 50
lever, 22, 42, 50
-levitin[fi], 33
leviäminen[fi], 31
levlo[se], 85
levlomašiidna[se], 85
levy[fi], 26, 63
levyjarru[fi], 26
levysakset[fi], 77, 90
levyseppä[fi], 90
lid, 11
liekkasárvu[se], 43
lieriöjyrsin[fi], 24
lieriöotsajyrsin[fi], 77
lievla[se], 85
lievlamašiidna[se], 85
light metal, 50
lihkadeapmi[se], 56
lihtdoaŋggat[se], 57
lihtolaš[se], 94
lihtti[se], 19
lihttu[se], 1
liibma[se], 40

liibmet[se], 40
liigeávnnas[se], 34
liigebuvtta[se], 11
liigenoađuheapmi[se], 60
liigeoassi[se], 82
liike[fi], 56
liima[fi], 40
liimata[fi], 40
liitin[fi], 1, 16, 21, 94
liitos[fi], 48
liittää[fi], 21
liitäntä[fi], 47
likerettar[nn], 69
likeretter[nb], 69
likestraum[nn], 25
likestrøm[nb], 25
likevekt[nb], 6
likriktare[sv], 69
likström[sv], 25
liktet[se], 64
lim[nb], 40
lim[sv], 40
lime[nb], 40
limiliitos[fi], 35
limit value, 50
limittäisliitos[fi], 35
limitys[fi], 60
limma[sv], 40
linda[sv], 98
lindning[sv], 98
lindningsvarv[sv], 93
line voltage, 52
lining, 16
linjal[nb], 72
linjal[sv], 72
linjála[se], 72
liquid, 50
lisäaine[fi], 34
liukua[fi], 79, 80
liukukytkin[fi], 79
liukulaakeri[fi], 80
liukuvastus[fi], 80
liuote[fi], 81
liuotinaine[fi], 81
live centre, 50
ljuddämpare[sv], 78, 79
ljusbåge[sv], 29
load, 51
loadbearing capacity, 12
loahppagieđahallan[se], 34
loaktinnannodat[se], 96
loaktinnanosvuohta[se], 96
loaktinstálli[se], 96
loaktu[se], 96
loaŋkkisteapmi[se], 6
location, 79
locking device, 51
locking ring, 51
locking washer, 51
locknut, 21
lodde[nb], 81
loddebolt[nb], 81
loddetinn[nb], 81
lodding[nb], 81
loddrett[nb], 61
lodrät[sv], 61
lohkkadan-[se], 75
lohkkadanrieggabasttat[se], 14
lohkkadanriekkis[se], 51
lohko[fi], 8
lohtegurraheavval[se], 83
lohtegurrajurssan[se], 48
lohteluodda[se], 48
lohteluoddaheavval[se], 83
lohteruopma[se], 18
lohtesveisa[se], 34
lohti[se], 97
lokke[nb], 67
loktenstággu[se], 50
loktenvávdna[se], 42
loop, 9
loose jaws, 51
loosen, 51
lossa[sv], 51
lotnolasbasttat[se], 17

lotnolasčoavdda[se], 17
lotnolasviŋkil[se], 94
lovtton[se], 21
low pressure, 51
lubricant, 51
lubricate, 51
lubricating film, 51
lubricating grease, 51
lubricating oil, 52
ludnemašiidna[se], 72
ludni[se], 24
luft(av)kjølt[nb], 3
luftfilter[nb], 3
luftfilter[sv], 3
lufthammar[nn], 3
lufthammare[sv], 3
luftkyld[sv], 3
luftpistol[nb], 2
luftpistol[sv], 2
lug, 78
luistaa[fi], 80
luistattaa[fi], 80
luisti[fi], 79
luistiventtiili[fi], 39
luja pakotustiukkuus[fi], 43
lukitsin[fi], 51
lukituslevy[fi], 75
lukitusrengas[fi], 51
lukka krets[nb], 16
lukka krins[nn], 16
lukkoaluslevy[fi], 51
lukkopihdit[fi], 40
lukkorengas[fi], 80
lukkorengaspihdit[fi], 14
lumisokeus[fi], 4
luodda[se], 80
luoddanibba[se], 37
luoitahat[se], 59
luokčat[se], 14
luondduviđá ávnnas[se], 69
luondduviđá olju[se], 22
luonnos[fi], 96
luođđa[se], 6
luođđadoalan[se], 6
luođđaláger[se], 6
luovččan[se], 14
luovusávju[se], 10
lutningsmätare[sv], 4
lutningsvinkel[sv], 68
luvda[se], 10
luvvadanávnnas[se], 81
luvvadat[se], 81
luvvadus[se], 30
luvvenlohti[se], 29
lyddempar[nn], 78, 79
lyddemper[nb], 78, 79
lyftkran[sv], 21
lyftögla[sv], 32
lyijyakku[fi], 49
lyijytön bensiini[fi], 49
lysboge[nn], 29
lysbue[nb], 29
lysnett[nb], 52
lyspære[nb], 10
läckage[sv], 50
läge[sv], 79
lämmönkestävä[fi], 43
lämpöarvo[fi], 43
lämpökäsittely[fi], 43
längdutvidgningskoefficient[sv], 16
länkiharppi[fi], 60
länkskruv[sv], 32
läppning[sv], 49
lære[nb], 40
lättmetall[sv], 50
lävistin[fi], 44
lävistyskone[fi], 67
lävistää[fi], 67
löda[sv], 81
lödkolv[sv], 81
lödning[sv], 81
lödtenn[sv], 81
løpekatt[nb], 92
løpekran[nb], 92
løs(n)e[nb], 51

løsegg[nb], 10
lösegg[sv], 10
løsemiddel[nb], 81
lösenord[sv], 60
lösning[sv], 30
lösningsmedel[sv], 81
løyse[nn], 51
løysemiddel[nb], 81
lågtryck[sv], 51
lågtrykk[nb], 51
långspån[sv], 93
låsbleck[sv], 75
låsbricka[sv], 51
låseblekk[nn], 75
låseblikk[nb], 75
låsering[nb], 51
låseringstang[nb], 14
låseringstong[nn], 14
låseskive[nb], 51
låsetang[nb], 40
låsetong[nn], 40
låsring[sv], 51

M

maadoittaa[fi], 28
maadoitus[fi], 29
maadoitusjohdin ukkosenjohdin[fi], 29
maajohdin[fi], 29
machine, 52
machine key, 97
machining, 35, 52
machining allowance, 99
magneetti[fi], 52
magneettikenttä[fi], 52
magneettisytytys[fi], 52
magneettivuo[fi], 37
magneettivuon tiheys[fi], 37
magnehta[se], 52
magnehtacahkkeheapmi[se], 52
magnehtagieddi[se], 52
magnet, 52
magnet[nb], 52
magnet[sv], 52
magnetfelt[nb], 52
magnetfält[sv], 52
magnetic field, 52
magnetisk felt[nb], 52
magnetoelectric ignition, 52
magnettenning[nb], 52
magnettändning[sv], 52
máhccanjođas[se], 70
máhccanrádji[se], 99
máhccanventiila[se], 13
máhccunmášiidna[se], 37
máhccut[se], 37
main cutting edge, 52
mains, 52
mains voltage, 52
maintenance, 52
major cutting edge, 52
mal[nb], 40
málbma[se], 59
málbmariggudus[se], 27
málgur[se], 27
mall[sv], 40
málle[se], 40, 55
malleable, 38
malleable cast iron, 52
mallet, 53, 79
malli[fi], 55
malline[fi], 40, 48
malm[nb], 59
malm[sv], 59
malmi[fi], 59
malmirikaste[fi], 27
mandrel stock, 43
manifold, 53
manifold[nb], 53
manifold[sv], 53
mannolat[se], 66
mannu[se], 19
manomehtar[se], 66
manometer, 66
manometer[nb], 66

manometer[sv], 66
manometri[fi], 66
manšeahtta[se], 79
mansjett[nb], 79
manual, 53
manuell[nb], 53
manuell[sv], 53
manufacturing plant, 32, 53
máŋggalođatovttastus[se], 82
maŋŋálaslakta[se], 76
maŋŋeáksil[se], 69
maŋŋegeašmohtor[se], 59
maŋŋejuvla[se], 69
margin, 90
marking gauge, 53, 86
markør[nb], 22
markör[sv], 22
mašineren[se], 52
mašineret[se], 52
maskinbearbetning[sv], 52
maskinere[nb], 52
maskinering[nb], 52
masomman[se], 8
masomn[nn], 8
masovn[nb], 8
massetetthet[nb], 24
massiv[nb], 81
massiv[sv], 81
masugn[sv], 8
masuuni[fi], 8
masuvdna[se], 8
mata[sv], 33
matalapaine[fi], 51
matarpump[sv], 33
matarspindel[sv], 33
mate[nb], 33
mateaksel[nb], 33
matehjul[nb], 68
matepumpe[nb], 33
materiaali[fi], 53
material, 53
material[sv], 53
materiale[nb], 53
mateskrue[nb], 33
mating[nb], 33
matning[sv], 33
matta[sv], 7
maul, 79
meaddilmannan[se], 11
meaddilventiila[se], 11
meanderiekkis[se], 62
meandestággu[se], 62
meandi[se], 62
mearka[se], 78
measure, 53
measure out, 25
measurement, 53
measuring, 53
measuring surface, 54
measuring tape, 54
measuring tool, 54
mechanic, 54
mechanical, 54
mechanician, 54
mechanism, 54
mechanization, 54
med-[nb], 50
medbringare[sv], 12
medbringarskiva[sv], 49
medbringer[nb], 12
medbringerskive[nb], 49
medfresing[nb], 16
medfräsning[sv], 16
medtakar[nn], 12
medtakarskive[nn], 49
mehtarlaš[se], 55
mehtarmihtádas[se], 46
meisel[nb], 14
meisle[nb], 14
meisset[se], 10
meistauskone[fi], 67
meisti[fi], 67
meistää[fi], 67
mejsel[sv], 14
mejsla[sv], 14
mekaaninen seos[fi], 55

mekaaninen[fi], 54
mekánalaš seaguhus[se], 55
mekánalaš[se], 54
mekanihkkalaš[se], 54
mekanihkkár[se], 54
mekanikar[nn], 54
mekaniker[nb], 54
mekaniker[sv], 54
mekaniseren[se], 54
mekanisering[nb], 54
mekanisering[sv], 54
mekanisk blanding[nb], 55
mekanisk blandning[sv], 55
mekanisk[nb], 54
mekanisk[sv], 54
mekanism[sv], 54
mekanisma[se], 54
mekanisme[nb], 54
mekanismi[fi], 54
mellanaxel[sv], 21
mellanlägg[sv], 78
mellanpassning[sv], 92
mellantapp[sv], 75
mellomaksel[nb], 21
mellomlegg[nb], 78
mellompasning[nb], 92
mellomtapp[nb], 75
melt, 54
melting point, 54
melting temperature, 54
melu[fi], 57
membran[nb], 54
membran[sv], 54
membrána[se], 54
membrane, 54
memory, 54
mend, 70
menu, 54
meny[nb], 54
meny[sv], 54
merkendurra[se], 13
merkenluovččan[se], 13
merkki[fi], 78
mesh, 30
messet[se], 10
messing[nb], 10
messinki[fi], 10
metal, 55
metal[sv], 55
metall[nb], 55
metálla[se], 55
metállaseaguhus[se], 3
metállašealgu[se], 55
metallglans[nb], 55
metallglans[sv], 55
metalli[fi], 55
metallic glint, 55
metallic lustre, 55
metallinkiilto[fi], 55
metalliseos[fi], 3
meterstokk[nb], 46
metric, 55
metrinen[fi], 55
metrisk[nb], 55
metrisk[sv], 55
mette[nb], 73
micro switch, 55
micrometer, 55
microprocessor, 55
midttapp[nb], 75
miehkki[se], 41
miehttejursan[se], 16
mielggas[se], 79
mierkavuoidi[se], 58
mihtádas[se], 54
mihtidanbáddi[se], 54
mihtidandiibmu[se], 25
mihtidanfárppal[se], 40
mihtidanneavvu[se], 54
mihtidannjunni[se], 19
mihtidanoalul[se], 54
mihtidanolggoš[se], 54
mihtideapmi[se], 53
mihttár[se], 54
mihttobidjan[se], 53
mihttodallat[se], 25

mihttoovttadat[se], 94
mikrobotkkán[se], 55
mikrobrytar[nn], 55
mikrobryter[nb], 55
mikrodoaimmár[se], 55
mikrokytkin[fi], 55
mikrometer[nb], 55
mikrometer[sv], 55
mikrometri[fi], 55
mikromihtádas[se], 55
mikroprocessor[sv], 55
mikroprosessor[nb], 55
mikroprosessori[fi], 55
mikroströmbrytare[sv], 55
mikrosuoritin[fi], 55
mill, 55
milling cutter, 55
milling machine, 55
minne[nb], 54
minne[sv], 54
minor cutting edge, 75
mitoittaa[fi], 25, 53
mitre rule, 8
mittaaminen[fi], 53
mittain[fi], 54
mittakaava[fi], 73
mittakello[fi], 25
mittakärki[fi], 19
mittanauha[fi], 54
mittapala[fi], 40
mittarumpu[fi], 40
mittaus[fi], 53
mittauspinta[fi], 54
mittaustyökalu[fi], 54
mittayksikkö[fi], 94
mixer tap, 55
mixing, 55
mixing battery, 55
mixing chamber, 55
mixing cock, 55
mixture, 55
mjuklodding[nb], 81
model, 55
modell[nb], 55
modell[sv], 55
modular measure, 56
mohtor[se], 56
mohtorgiddehat[se], 30
mohtorgielka[se], 80
mohtorgorut[se], 23
mohtorliekkan[se], 31
mohtorsahá[se], 65
mohtorsihkkal[se], 56
mohtorsuodjebotkkon[se], 56
molecule, 56
molekyl[nb], 56
molekyl[sv], 56
molekyla[se], 56
molekyyli[fi], 56
molsa[se], 40
molssačoavdda[se], 2
molssarávdnji[se], 3
molssohahtti galján[se], 2
molsson[se], 40
molssonkássa[se], 40
molsut[se], 13
molten mass, 54
molten pool, 56
momeanta[se], 91
momeantačoavdda[se], 91
moment, 91
moment[nb], 91
moment[sv], 91
momentnyckel[sv], 91
momentnøkkel[nb], 91
momentti[fi], 91
momenttiavain[fi], 91
moniotepihdit[fi], 96
monitor, 74
-monn[nb], 99
montera[sv], 5
montere[nb], 5
monteret[se], 5
moottori[fi], 56
moottorikelkka[fi], 80
moottorilämmitin[fi], 31

moottorin kiinnitin[fi], 30
moottorin suojakytkin[fi], 56
moottoripolkupyörä[fi], 56
moottoripyörä[fi], 56
moottorisaha[fi], 65
moped, 56
moped[nb], 56
moped[sv], 56
mopeda[se], 56
mopo[fi], 56
motfresing[nb], 20
motfräsning[sv], 20
motion, 56
motive force, 56
motive power, 56
motor, 56
motor cycle, 56
motor protecting switch, 56
motor)[nb], 84
motor[nb], 56
motor[sv], 56
motorbike, 56
motorblokk[nb], 23
motor-car, 12
motorcykel[sv], 56
motorfeste[nb], 30
motorfäste[sv], 30
motorsag[nb], 65
motorskydd[sv], 56
motorsykkel[nb], 56
motorsåg[sv], 65
motorvarmar[nn], 31
motorvarmer[nb], 31
motorvernbrytar[nn], 56
motorvernbryter[nb], 56
motorvärmare[sv], 31
motstand[nb], 70
motstandsevne[nb], 70
motstånd[sv], 70
motståndskraft[sv], 70
motsveising[nb], 71
motsvetsning[sv], 50, 71
motvekt[nb], 6
motvikt[sv], 6
moukari[fi], 79
mould, 12, 56
mount, 5
mountings, 35
mouse, 56
mouth piece, 57
movement, 56
mower, 57
mowing machine, 57
muddenbelle[se], 36
muddosirdu[se], 61
muddu[se], 61
muff[sv], 80
muffa[se], 80
muffe[nb], 80
muffler, 78, 79
muhkebiđggon[se], 33
muhtán[se], 20
muhtter[se], 58
muhvi[fi], 80
muisti[fi], 54
muitu[se], 54
multipower end cutting nippers, 9
munnstykke[nb], 57
munstycke[sv], 57, 96
muodonmuutos[fi], 24
muorraskruvva[se], 99
muotava[fi], 24
muotti[fi], 56
muovaava työstö[fi], 38
muovailtava[fi], 28
muovaus[fi], 38
muovautuva[fi], 28
muovi[fi], 63
muovivasara[fi], 63
muovveveažir[se], 63
muovvi[se], 63
murdinmolsson[se], 70
murtolastu[fi], 87
murtopinta[fi], 87
mus[nb], 56

mus[sv], 56
mutter[nb], 58
mutter[sv], 58
mutteri[fi], 58
muttertrekker[nb], 3
muunnin[fi], 95
muuntaja[fi], 20, 92
myötähitsaus[fi], 50
myötäjyrsintä[fi], 16
männän isku[fi], 86
männänrengas[fi], 62
männänvarsi[fi], 62
mäntä[fi], 62
mässing[sv], 10
mätklocka[sv], 25
mätning[sv], 53
mätspets[sv], 19
mätstock[sv], 46
mätta[sv], 73
mättrumma[sv], 40
mätverktyg[sv], 54
mätyta[sv], 54
mönkijä[fi], 38
-målar[nn], 71
måleband[nb], 54
måleining[nn], 94
målekjeft[nb], 54
målenhet[nb], 94
målespiss[nb], 19
måletrommel[nb], 40
måleur[nb], 25
måleverktøy[nb], 54
måling[nb], 53
målsetje[nn], 53
målsette[nb], 53
måttenhet[sv], 94
måttsätta[sv], 53

N

nábár[se], 27
nagle[nb], 71
náhpi[se], 64
náhpul[se], 60
nail, 57
nail puller, 57
nákca[se], 11
nakutuksenkestävyys[fi], 4
nállolager[se], 57
nammaoassi[se], 90
nannen[se], 5
nannenstálli[se], 70
nannodat[se], 86
nanosmahtti[se], 5
nanosmahttinstálli[se], 70
nanosmahttit[se], 86
nađđa[se], 42
nađđadit[se], 42
napa[fi], 45, 64
nárbma[se], 53
náskál[se], 5
naskali[fi], 5
nástečoavdda[se], 81
nástegolbmačiegatgoallus[se], 84
násterabasčoavdda[se], 17
násteskruvvaruovdi[se], 22
naudstopp[nn], 30
naula[fi], 57
naulanvedin[fi], 57
nav[nb], 45
nav[sv], 45
nave, 45
návli[se], 71
neavvobumbá[se], 91
neavvogiddehat[se], 91
neavvojohtolat[se], 91
neavvorávdi[se], 91
nebbtang[nb], 36
nebbtong[nn], 36
needle bearing, 57
negative pressure, 51
nelitahtimoottori[fi], 38
neliökantaruuvi[fi], 83
neseradius[nb], 20
neste[fi], 50
neste-[fi], 45

nettspenning[nb], 52
neulalaakeri[fi], 57
nihppel[se], 57
niibestálli[se], 49
niibi[se], 49
niitata[fi], 71
niitti[fi], 71
niittokone[fi], 57
niittosilppuri[fi], 37
nikkarinsaha[fi], 48
nippa[fi], 57
nippel[nb], 57
nippel[sv], 57
nippeli[fi], 57
nippers, 57
nipple, 57
nit[sv], 71
nita[sv], 71
nittång[sv], 71
nivel[fi], 12
nivelavain[fi], 87
nivelmitta[fi], 46
nivelväännin[fi], 87
njalbi[se], 50
njálbmebihttá[se], 57
njalpagoallus[se], 79
njalpalavtta[se], 79
njalpit[se], 80
njaman[se], 37
njamman[se], 86
njárbbadas[se], 89
njealjeborat skruvva[se], 83
njealječiegat skruvva[se], 83
njealječiegatjeaŋga[se], 36
njealjedávttatmohtor[se], 38
njealjejuvllat (sihkkal)[se], 38
njielahat[se], 59
njoalveláger[se], 80
njulgen[se], 3
njulgenneavvu[se], 85
njulggonas[se], 63
njunneáksil[se], 11
njunnebasttat[se], 36
njunneradius[se], 20
njuolggan[se], 69
njuolggogidden[se], 78
njuolggolaktin[se], 78
njuolggolinjála[se], 72
no load voltage, 58
noađuheapmi[se], 51
noađuhit[se], 51
noavki[se], 94
noble gas, 57
nodular cast iron, 57
noeta[fi], 81
noggrannhet[sv], 65
noise, 57
noki[fi], 81
nokka-akseli[fi], 11
nollakohta[fi], 99
nollapiste[fi], 99
nolläge[sv], 99
no-load, 45
non-acceptance limit, 58
nonie[nb], 95
nonie[se], 95
nonie[sv], 95
nonieskala[sv], 95
nonius[nb], 95
noniusa[se], 95
non-return valve, 13
noonio[fi], 95
nopeus[fi], 82
nopeusmittari[fi], 82
normaaliasento[fi], 58
normal position, 58
normálasajádat[se], 58
normalisera[sv], 58
normalisere[nb], 58
normaliseret[se], 58
normalisoida[fi], 58
normalize, 58
normalläge[sv], 58
normalstilling[nb], 58
nosradie[sv], 20
nostosilmukka[fi], 32

nostovaunu[fi], 42, 92
nosturi[fi], 21, 47, 98
notch, 22
notkkaldat[se], 34
nozzle, 57, 58
nuija[fi], 53
nullačuokkis[se], 99
nullpunkt[nb], 99
number of teeth, 58
numeerinen[fi], 25
nuorrahuvvat[se], 7
nuorrašuvvat[se], 7
nuorrutettu teräs[fi], 89
nuorrutus[fi], 89
nuoskkideapmi[se], 64
nuppáldas[se], 75
nut, 58
nyckelgap[sv], 98
nyckelvidd[sv], 98
näppäimistö[fi], 48
nätspänning[sv], 52
näyttö[fi], 74
nødstopp[nb], 30
nödstopp[sv], 30
nøkkelvidd[nn], 98
nøkkelvidde[nb], 98
nötning[sv], 96
nötningsmotstånd[sv], 96
nøyaktighet[nb], 65
nøytralstilling[nb], 58
nålelager[nb], 57
nållager[sv], 57

O

oaiveávju[se], 52
oalásruovdi[se], 96
oalgeávju[se], 75
oalul[se], 48
oalulgiddehat[se], 47
oasselávgu[se], 74
oasselistu[se], 60
obalans[sv], 94
octane number, 58
ohcu[se], 26
ohennin[fi], 89
ohitusventtiili[fi], 11
ohivirtaus[fi], 11
ohjain[fi], 20, 48
ohjaus[fi], 85
ohjausosa[fi], 20
ohjauspyörä[fi], 42
ohjausrauta[fi], 96
ohjaustekniikka[fi], 20
ohjausteline[fi], 37
ohjelma[fi], 66
ohjelmoida[fi], 66
oikaisu[fi], 3
oikaisulaite[fi], 63
oikaisutyökalut[fi], 85
oikosulku[fi], 78
oikotaso[fi], 87
oil, 58
oil bath, 58
oil can, 58
oil groove, 58
oil mist lubricator, 58
oktaaniluku[fi], 58
oktánalohku[se], 58
oktantal[nn], 58
oktantal[sv], 58
oktantall[nb], 58
oktávuođaguovki[se], 19
oktiibidjan[se], 48
oktiigidden[se], 48
olake[fi], 78
olasrauta[fi], 96
olboge[nn], 29
olggoš[se], 86
olggošgieđahallan[se], 87
olggošmálle[se], 87
olggosoassi[se], 59
olggošroamši[se], 87
olggošromšodat[se], 87
olgoriekkis[se], 59
olja[sv], 58
olje[nb], 58

oljebad[nb], 58
oljebad[sv], 58
oljedimbildare[sv], 58
oljekanna[sv], 58
oljekanne[nb], 58
oljesump[nb], 58
oljogádnu[se], 58
oljolávgu[se], 58
olju[se], 58
omdreiing[nb], 70
omdreiingsmåler[nb], 71
omdreiingstall[nb], 72
omformar[nn], 20
-omformar[nn], 97
omformare[sv], 20
omformer[nb], 20
ominais-[fi], 82
omløp[nb], 11
omløpsventil[nb], 11
open circuit voltage, 58
open ringnøkkel[nn], 36
open-end spanner, 59
open-end wrench, 59
operating instructions, 53
operation, 59, 85
-opning[nn], 19
oppalašjursanmašiidna[se], 94
oppladbar[nb], 69
oppløsning[nb], 30
oppløysing[nb], 30
oppretting[nb], 3
opprettingsverktøy[nb], 85
opprike[nn], 31
oppriking[nn], 31
opprømme[nb], 69
oppspenningsdor[nb], 15
oppvarme[nb], 43
ore, 59
ore concentrate, 27
O-rengas[fi], 59
orepinne[nb], 80
O-riekkis[se], 59
O-ring seal, 59
O-ring[nb], 59
O-ring[sv], 59
osaluettelo[fi], 60
oscillate, 97
oscilloscope, 59
oscilloskohppa[se], 59
oscilloskop[nb], 59
oscilloskop[sv], 59
osienpesulaite[fi], 74
oskarp[sv], 8
oskilloskooppi[fi], 59
osoitin[fi], 46
otsajyrsin[fi], 32
outboard engine, 59
outer race, 59
outer ring, 59
outlet, 37, 59
outlet tube, 26
output unit, 59
outside calipers, 60
ovdaáksil[se], 39
ovdagealdda[se], 66
ovdajuvlageassin[se], 39
ovdastelle[se], 37
overbelastning[nb], 60
overflate[nb], 86
overflatebehandling[nb], 87
overflatenormal[nb], 87
overflateruhet[nb], 87
overflateruleik[nn], 87
overføring[nb], 92
overhead travelling crane, 92
over-lapping, 60
overlappskjøt[nb], 60
overlappskøyt[nn], 60
overload, 60
overpressure, 60
overtrykk[nb], 60
ovttadoaimmat sylinddar[se], 79
oxide scale, 73

P

p. control valve, 66
paalain[fi], 43
packing, 39, 75
packning[sv], 39
paikka[fi], 79
painaa[fi], 65
paineenalennusventtiili[fi], 66, 69
paineensäädin[fi], 66
paineilma[fi], 17
paineilma-[fi], 64
paineilmapistooli[fi], 2
paineilmasuutin[fi], 2
paineilmavasara[fi], 3
painemittari[fi], 66
painepesuri[fi], 44
painike[fi], 67
pair of compasses, 60
paisuntakuoripultti[fi], 32
paja[fi], 80
pajahitsaus[fi], 38
pajapihdit[fi], 38
pajavasara[fi], 79
pakning[nb], 39
pakningsskjerar[nn], 2
pakningsskjærer[nb], 2
pakokaasu[fi], 31
pakokaasumittari[fi], 31
pakoputki[fi], 31
pakoputkisto[fi], 31
pakosarja[fi], 53
pakotustiukkuus[fi], 90
palava[fi], 17
palhjul[nb], 68
palje[fi], 7
palkki[fi], 7
pall[nb], 70
pall[sv], 70
palla[se], 70
palle[nb], 70
pallografiittirauta[fi], 57
pallomainen laakeri[fi], 82
pallomainen[fi], 82
paloarvo[fi], 43
paloletku[fi], 35
palstasovite[fi], 35
paluujohto[fi], 70
panel[nb], 35
paralleallabihttá[se], 40
paralleallagoallus[se], 78
parallel connection, 78
parallel interface, 60
parallell gränssnitt[sv], 60
parallellbit[sv], 40
parallellkloss[nb], 40
parallellkopling[nb], 78
parallellkoppling[sv], 78
parallelt grensesnitt[nb], 60
paristo[fi], 7
particle, 60
partihkal[se], 60
partikel[sv], 60
partikkel[nb], 60
parts list, 60
pasning[nb], 35
passar[nn], 26
passare[sv], 26
passbit[nb], 40
passbit[sv], 40
passer[nb], 26
passning[sv], 35
passord[nb], 60
password, 60
peannaveažir[se], 85
pedal, 60
pedal[nb], 60
pedal[sv], 60
peg, 60
pelarborrmaskin[sv], 94
pelti[fi], 77
peltisakset[fi], 90
peltiseppä[fi], 90
pendel[nb], 60
pendel[sv], 60
pendla[sv], 97

pendle[nb], 97
pendulum, 60
penhammare[sv], 85
pennhammar[nn], 85
pennhammer[nb], 85
percolate, 60
percussion drilling machine, 42
perforated plate, 61
perforerad platta[sv], 61
perforert plate[nb], 61
period, 23
period[sv], 23
periode[nb], 23
periodical, 61
periodisk[nb], 61
periodisk[sv], 61
permanent, 61
permanent[nb], 61
permanent[sv], 61
perpendicular, 61
perusmitta[fi], 56
perustus[fi], 38
peruutusvaihde[fi], 70
perämoottori[fi], 59
petrol, 39
petrol engine, 61
phase, 61
phase shift, 61
phase wear, 61
pidin[fi], 44
pidätinruuvi[fi], 76
pienahitsi[fi], 34
pieni vetopyörä[fi], 61
pig iron, 61
pihdit[fi], 91
piirrottaa[fi], 74
piirrotus[fi], 74
piirrotuskelkka[fi], 86
piirtokärki[fi], 75
piirtopuikko[fi], 75
piirturi[fi], 64
piirustus[fi], 27
piirustuslauta[fi], 27
pikateräs[fi], 44
pillar drilling machine, 94
pilot tap, 88
pin, 60, 61, 83
pin hammer, 85
pincers, 57
pincett[sv], 62
pinion, 61
pinjong[nb], 61
pinjong[sv], 61
pinjoŋga[se], 61
pinnankarheus[fi], 87
pinnankarheusmalli[fi], 87
pinne[nb], 61
pinne[sv], 61
pinnefres[nb], 30
pinnfräs[sv], 30
pinnoite[fi], 16
pinol[nb], 88
pinol[sv], 88
pinolbohcci[se], 88
pinoldocka[sv], 88
pinoldokke[nb], 88
pinolnjunni[se], 88
pinolrør[nb], 88
pinolspiss[nb], 88
pinseahtat[se], 62
pinsetit[fi], 62
pinsett[nb], 62
pinta[fi], 86
pintakarkaisu[fi], 12
pintakäsittely[fi], 87
pinzers, 62
pipe, 62
pipe cutter, 62
pipe thread, 62
pipe tongs, 62
pipe wrench, 62
pipe[nb], 9
pipenøkkel[nb], 9
pistehitsaus[fi], 83
pistehitsauskone[fi], 83

pistepuikko[fi], 13
pistin[fi], 80
pistomitta[fi], 89
piston, 62
piston ring, 62
piston rod, 62
piston stroke, 86
piston travel, 50
pistorasia[fi], 80
pistosaha[fi], 89
pistoterä[fi], 80
pistotulppa[fi], 64
pistää[fi], 22
pitch circle, 62
pitch circle of bolt holes, 62
pitch of thread, 74
pitch rack, 68
pitting wear, 62
pituuslaajenemiskerroin[fi], 16
pivot, 41
pláhtta[se], 63
pláhttaskárrit[se], 77
pláhttaskierat[se], 77
pláhttaviŋkil[se], 83
plain bearing, 80
plain milling cutter, 24
plain milling machine, 45
plan[nb], 36
plan[sv], 36
plana[sv], 63
planbord[nb], 87
plandreie[nb], 32
plane, 63
plane[nb], 63
planer, 63
planfres[nb], 32
planfräs[sv], 32
planfräsmaskin[sv], 45
planing bench, 12
planing machine, 63
planing tool, 63
planskiva[sv], 32
planskive[nb], 32
planslipemaskin[nb], 86
planslipmaskin[sv], 86
plansvarva[sv], 32
plant, 63
plasma cutter, 63
plasmačuohpan[se], 63
plasmaleikkain[fi], 63
plasmaleikkauslaite[fi], 63
plasmaskjerar[nn], 63
plasmaskjærer[nb], 63
plasmaskärare[sv], 63
plast[nb], 63
plast[sv], 63
plásta[se], 63
plástalaš gieđahallan[se], 38
plástalaš[se], 28
plástaveažir[se], 63
plasthammar[nn], 63
plasthammare[sv], 63
plasthammer[nb], 63
plastic, 28
plastic mallet, 63
plastics, 63
plastisk bearbeiding[nb], 38
plastisk bearbetning[sv], 38
plastisk[nb], 28
plastisk[sv], 28
plate, 26, 63
plate shears, 77
plate spring, 50
plate[nb], 63
platelager[nb], 42
platesaks[nb], 77
platevinkel[nb], 83
platinastift[nb], 19
platta[sv], 63
plattstål[sv], 36
pliers, 91
plog[nb], 64
plog[sv], 64
plottar[nn], 64
plotter, 64

plotter[nb], 64
plough, 64
plug, 60, 64
plug tap, 64, 75
pluga[se], 64
plugg[nb], 60, 64
plugg[sv], 60
pluggačoavdda[se], 82
plugghette[nb], 82
pluggledning[nb], 64
pluggleidning[nn], 64
pluggnøkkel[nb], 82
plugwire, 64
plumber, 64
plunger, 64
plåt[sv], 63, 77
plåtsax[sv], 77, 90
plåtslagare[sv], 90
pneumaattinen[fi], 64
pneumáhtalaš[se], 64
pneumatic, 64
pneumatic hammer, 3
pneumatisk[nb], 64
pneumatisk[sv], 64
poikittaisluisti[fi], 22
poikittaispalkki[fi], 93
poikkema[fi], 25
poikkileikkaus[fi], 22
poikkiparru[fi], 93
poikkisärmä[fi], 49
poikkitaiskelkka[fi], 22
poistaa jäysteet[fi], 15
poistoimuri[fi], 37
pol[nb], 64
pol[sv], 64
pole, 64
polera[sv], 64
polere[nb], 64
polish, 64
poljin[fi], 60
pollution, 64
polttaminen[fi], 17
polttoaine[fi], 39
polttomoottori[fi], 17
pop rivet, 64
popnagle[nb], 64
popnagletangb blindnagletang[nb], 71
popnávledoaŋggat[se], 71
popnávli[se], 64
pop-niitti[fi], 64
pop-niittipihdit[fi], 71
popnit[sv], 64
por[sv], 65
pora[fi], 27
porakone[fi], 27
porata[fi], 27
porbildning[sv], 65
pore, 65
pore forming, 65
pore[nb], 65
poredanning[nb], 65
porous, 65
portaaton[fi], 85
porøs[nb], 65
porös[sv], 65
posisjon[nn], 79
positive pressure, 60
potensiell energi[nb], 65
potentiaalienergia[fi], 65
potential energy, 65
potentiell energi[sv], 65
potkuri[fi], 66
powder, 65
power drive, 65
power grid, 52
power network, 52
power saw, 65
power source, 65
power stroke, 65
power transmission, 65
precision, 65
precision[sv], 65
prepare, 65
presisjon[nb], 65
press, 65, 66

press[sv], 66
pressa[sv], 65
presse[nb], 65, 66
pressmonn[nb], 35
pressmån[sv], 35
presspasning[nb], 43
presspassning[sv], 43
pressure gauge, 66
pressure governor, 66
pressure reducing valve, 66
pressure regulator, 66
pre-stressing, 66
pre-tensioning, 66
primára[se], 66
primary, 66
primary circuit, 66
primær[nb], 66
primär[sv], 66
primærkrets[nb], 66
primärkrets[sv], 66
primærkrins[nn], 66
printer, 66
prisma[sv], 7
probe, 24
process, 65, 66
process[sv], 66
processor, 66
processor[sv], 66
program, 66
program[nb], 66
program[sv], 66
programera[sv], 66
prográmma[se], 66
programme, 66
programmere[nb], 66
prográmmeret[se], 66
pronssi[fi], 10
propealla[se], 66
propell[nb], 66
propeller, 66
propeller[sv], 66
proportion of ingredients, 67
proseassa[se], 66
prosess[nb], 66
prosessi[fi], 66
prosessor[nb], 66
prosessori[fi], 66
protective equipment, 67
protective helmet, 67
protective spectacles, 67
prov[sv], 89
prøve[nb], 89
puhallin[fi], 33
puhdistaa[fi], 64
puller, 67
pulley, 67, 72
pulley block, 8
puls[nb], 67
puls[sv], 67
pulsating, 61
pulse, 67
pulserande[nn], 61
pulserende[nb], 61
pulssi[fi], 67
pultti[fi], 9
pulttisakset[fi], 9
pulver[nb], 65
pulver[sv], 65
pulveri[fi], 65
pump, 67
pump[sv], 67
pumpa[sv], 67
pumpata[fi], 67
pumpe[nb], 67
pumppu[fi], 67
punch, 13, 27, 67
punching machine, 67
punching tool, 67
punktsveiseapparat[nb], 83
punktsveising[nb], 83
punktsvetsning[sv], 83
punktsvetsningsmaskin[sv], 83
puolijohde[fi], 76
puolipyöreä viila[fi], 42
puolivalmiste[fi], 42

puristaa kokoon[fi], 17
puristaa[fi], 65
puristin[fi], 66
puristus[fi], 17
puristusholkki[fi], 1
puristuskone[fi], 66
puristuspainemittari[fi], 17
puristusrauta[fi], 15
puristustiukkuus[fi], 43
purkaa[fi], 26
purse[fi], 10
push button, 67
puskuri[fi], 10
pusse[nb], 64
putken mutka[fi], 29
putkenkatkaisin[fi], 62
putki[fi], 62
putkiasentaja[fi], 64
putkikierre[fi], 62
putkileikkuri[fi], 62
putkipihdit[fi], 62
putkisokka[fi], 64
putsa[sv], 64
putty, 68
puukko[fi], 77
puuruuvi[fi], 99
puuseppä[fi], 12, 48
pykälä[fi], 49
pylväsporakone[fi], 94
pystysuora[fi], 61
pysyvyys[fi], 84
pysyvä[fi], 84
pyälletty tappi[fi], 37
pyällyskehrä[fi], 76
pyällystyökalu[fi], 76
pyältää[fi], 76
pyörimisnopeus[fi], 72
pyörimisnopeusmittari[fi], 71
pyörivä keskiökärki[fi], 50
pyöriä[fi], 72
pyörteisyys[fi], 93
pyörä[fi], 98
pyöröhiomakone[fi], 15
pyörökärkipihdit[fi], 72
pyörösaha[fi], 15
pyöröviila[fi], 72
päittäislaakeri[fi], 90
pää-[fi], 66
pääleikkuusärmä[fi], 52
päällehitsaus[fi], 15
pääpiiri[fi], 66
päästäminen[fi], 4
päästää[fi], 4
päästö[fi], 4
päästökarkaistu teräs[fi], 89
päästökarkaisu[fi], 89
päästökulma[fi], 5, 15
päästöpinta[fi], 15
pölynimuri[fi], 94
påhengsmotor[nb], 59
påleggssveising[nb], 15
påsvetsning[sv], 15

Q

quenching, 68

R

raaka-aine[fi], 69
raakaöljy[fi], 22
rabas čoavdda[se], 59
rabas gierdočoavdda[se], 36
rabas nástečoavdda[se], 36
rabbet, 37, 68
race, 68
rack, 68, 84
radiaalinen[fi], 68
radiáhtor[se], 68
radial, 68
radial bearing, 68
radial drilling machine, 68
radial[nb], 68
radial[sv], 68
radiála[se], 68
radialboremaskin[nb], 68
radialborrmaskin[sv], 68
radialbovrenmašiidna[se], 68

radiallager[nb], 68
radialláger[se], 68
radiallager[sv], 68
radiator, 68
radiator[nb], 68
radiell[nb], 68
radiell[sv], 68
rádjemihttu[se], 50
rádna[se], 27
rádnet[se], 27
raekoko[fi], 40
rággu[se], 71
ráhkadus[se], 24, 63
ráhkadusmašiidna[se], 19
ráhkadusstálli[se], 19
ráhkku[se], 65
ráhtta[se], 42
ráhtte[se], 39
ráidogoallus[se], 76
ráiganbasttat[se], 67
ráiganbiipu[se], 44
ráigegierdu[se], 62
ráigeguoskkahat[se], 80
ráiggaspláhtta[se], 61
railo[fi], 41
railon valmistus[fi], 41
rajamitta[fi], 50
rake, 68
-rakeinen[fi], 34
rakenne[fi], 24
rakenneteräs[fi], 19
rakennus[fi], 19, 63
rakomitta[fi], 33
rakotulkki[fi], 33
ránesbákti[se], 77
-ražaheapmi[se], 60
rasp, 16
rasp[nb], 16
rasp[sv], 16
ráspa[se], 16
raspi[fi], 16
rassehit[se], 68
rasvata[fi], 51
ratchet handle, 68
ratchet wheel, 68
ratchet wrench, 68
rate, 82
ratio of mixture, 67
ratt[nb], 42
ratt[sv], 42
ráttis[se], 98
raudoitus[fi], 35
rauta[fi], 47
rautakanki[fi], 22
rautamalmi[fi], 47
rautasaha[fi], 42
ravdadeapmi[se], 35
ravdadit[se], 8
ravdaloaktu[se], 61
ravdet[se], 8
rávdi[se], 80
rávdnje-aggregáhta[se], 2
rávdnjebiire[se], 30
rávdnjegáldu[se], 65
rávdnjehálti[se], 25
rávdnjejuogan[se], 26
rávdnji[se], 22
ravkečuovga[se], 47
ravki[se], 61
ravkkas[se], 67
raw material, 69
reaidu[se], 91
reakta[se], 39
ream, 69
reamer, 69
reaming bit, 69
reaper, 57
rear axle, 69
rear wheel, 69
rebate, 37
rechargeable, 69
recoil, 48
record, 73
rectifier, 69
reduction valve, 69
reduksjonsventil[nb], 69

reel, 16, 69
register, 69
register belt, 69
register chain, 69
register[nb], 69
register[sv], 69
registerkedja[sv], 69
registerkjede[nb], 69
registerreim[nb], 69
registerrem[sv], 69
registtar[se], 69
registtarruopma[se], 69
registtarviđji[se], 69
reglerenteknihkka[se], 19
reglering[sv], 40
reglerteknik[sv], 19
regneenhet[nb], 66
reguláhtor[se], 70
regulation, 40
regulator, 70
regulator[nb], 70
regulator[sv], 70
reguleren[se], 40
regulering[nb], 40
reguleringsteknikk[nb], 19
reie[nb], 72
rei'itetty levy[fi], 61
reikälevy[fi], 61
reikämeisti[fi], 44
reikäpihdit[fi], 67
reim[nb], 27
reimdrift[nb], 7
reimskive[nb], 67
reinforce, 86
reinforcement, 86
reinforcement steel, 70
reinforcing, 5
rekisteri[fi], 69
rekneeining[nn], 66
relay, 70
rele[fi], 70
rele[nb], 70
rele[se], 70
relief angle, 15
relä[sv], 70
rem[sv], 27
remdrift[sv], 7
remote control, 70
remove burrs, 15
remskiva[sv], 67
rengas[fi], 94
rengasavain[fi], 81
repair, 70
reparasjon[nb], 70
reparation[sv], 70
reparatør[nb], 54
reparera[sv], 70
reparere[nb], 70
reservdel[sv], 82
reservedel[nb], 82
reservoir, 19
resistance, 70
resistance power, 70
resistans[nb], 70
resistánsa[se], 70
restore, 70
retning[nb], 25
retningslys[nb], 47
retningsventil[nb], 25
rettholt[nb], 72
returledning[nb], 70
returledning[sv], 70
returleidning[nn], 70
return line, 70
reusable pallet, 70
revers[nb], 70
reverse, 70
revolution, 70
revolution counter, 71
revolving punch pliers, 67
revre[se], 62
rib, 71
ribba[sv], 71
ribbe[nb], 71
rifle, 76
rifle[nb], 49

riflepinne[nb], 37
riggudeapmi[se], 31
riggudit[se], 31
riggudus[se], 27
rightward welding, 71
rigid ball bearing, 92
rigidity, 85
rihkkosággi[se], 37
rihkku[se], 49
rihkkuhit[se], 76
rihkon[se], 76
rihlaus[fi], 82
rihloittaa[fi], 76
rikastaa[fi], 31
rikastaminen[fi], 31
rikaste[fi], 27
rikastin[fi], 14
rikastus[fi], 31
rikki[fi], 86
riktning[sv], 3, 25
riktningsventil[sv], 25
riktplatta[sv], 87
riktverktyg[sv], 85
rilla[sv], 49
rille[nb], 49
rim, 71
rima[fi], 71
ringnyckel[sv], 81
ringnøkkel[nb], 81
rinnakkaiskytkentä[fi], 78
rinnakkaisliitäntä[fi], 60
rinnankytkentä[fi], 78
rinsing nozzle, 96
rintakulma[fi], 6
rintapinta[fi], 6
ripa[fi], 71
ripe[nb], 49, 74
ripple, 49
riss[nb], 74, 96
riššа[se], 86
risse[nb], 74
rissefot[nb], 86
rissenål[nb], 75
rissmål[nb], 53
ristikkoavain[fi], 98
ristikärkitaltta[fi], 22
ristitaltta[fi], 22
ritbräde[sv], 27
ritning[sv], 27
rits[sv], 74
ritsa[sv], 74
ritskubb[sv], 86
ritsmått[sv], 53
ritsnål[sv], 75
rivet, 71
rivet tool, 71
roahkkečoavdda[se], 44
roahkkohus[se], 30
roahtačoaskudeapmi[se], 68
roaisteruovdi[se], 74
roamši[se], 72
roaŋkeskruvvaruovdi[se], 4
roavvafiilu[se], 16
roavvagieđahallan[se], 72
roavvagortnat[se], 16
roavvastálli[se], 72
rod, 71
rohtor[se], 72
roll, 24
roller, 24, 71
roller bearing, 71
rolling bearing, 72
rolling machine, 72
rolling mill, 72
romšodat[se], 72
romurauta[fi], 74
roottori[fi], 72
roskalaatikko[fi], 39
roskasäiliö[fi], 39
rost[sv], 73
rosta[sv], 73
rostfritt stål[sv], 84
rostlösare[sv], 73
rotate, 72
rotational speed, 72
rotera[sv], 72

roterande dubb[sv], 50
roterande senterspiss[nn], 50
rotere[nb], 72
roterende senterspiss[nb], 50
rotor, 72
rotor[nb], 72
rotor[sv], 72
rotterumpe[nb], 89
rough edge, 10
rough-finish, 72
roughing, 72
roughing tool, 72
roughness, 72
rouhinta[fi], 72
rouhintaterä[fi], 72
rouhintaviila[fi], 16
round end pliers, 72
round file, 72
rovvi[se], 96
rowel, 72
rpm, 72
rubber, 72
rubbing, 72
ruhet[nb], 72
ruiskutus[fi], 46
ruiskutusventtiili[fi], 46
rule, 72
ruleik[nn], 72
ruler, 72
rull[nb], 71
rulla[fi], 71
rulla[se], 71
rullager[sv], 71
rullalaakeri[fi], 71
rullaláger[se], 71
rulle[sv], 71
rullelager[nb], 71
rullingslager[nb], 72
rullning[sv], 98
rullningscirkel[sv], 62
rullningslager[sv], 72
rumpu[fi], 28
rumpujarru[fi], 28
rundbalspress[sv], 43
rundfil[nb], 72
rundfil[sv], 72
rundskärare[sv], 2
rundslipemaskin[nb], 15
rundslipmaskin[sv], 15
rundtang[nb], 72
rundtong[nn], 72
rundtång[sv], 72
runko[fi], 7, 84
running, 59
run-out, 29
ruopma[se], 27
ruopmadoaibma[se], 7
ruopmaskearru[se], 67
ruossa[se], 98
ruossaluovččan[se], 22
ruosta[se], 73
ruostaluvvi[se], 73
ruostameahttun stálli[se], 84
ruoste[fi], 73
ruosteenirroitin[fi], 73
ruostua[fi], 73
ruostumaton teräs[fi], 84
ruostut[se], 73
ruoto[fi], 88
ruovdegákkan[se], 22
ruovdegáŋga[se], 22
ruovdemálbma[se], 47
ruovdesahá[se], 42
ruovdi[se], 47
rupture disc, 73
rusa[sv], 68
ruse[nb], 68
ruskalihtti[se], 39
rust, 73
rust killer, 73
rust removing fluid, 73
rust[nb], 73
ruste[nb], 73
rustfritt stål[nb], 84
rustløser[nb], 73
rustløysar[nn], 73

rusttet[se], 4
ruuvi[fi], 74
ruuviavain[fi], 81
ruuvikuljetin[fi], 33
ruuvimeisseli[fi], 74
ruuvin ulosvetäjä[fi], 74
ruuvipenkki[fi], 7
ruuvipuristin[fi], 15, 34
ruuvitaltta[fi], 74
ruvdnomuhtter[se], 12
ruvven[se], 72
ryntäyttää[fi], 68
räffla[sv], 76
räikkä[fi], 68
räikkäväännin[fi], 68
räjähdys[fi], 32
rätskiva[sv], 72
rømmer[nb], 69
rør[nb], 62
rör[sv], 62
rörelse[sv], 56
rørgjenge[nb], 62
rörgänga[sv], 62
rörkapare[sv], 62
rørkne[nb], 29
rörkrök[sv], 29
rørkuttar[nn], 62
rørkutter[nb], 62
rørlegger[nb], 64
rörläggare[sv], 64
rørsle[nn], 56
rørsplint[nb], 64
rørtang[nb], 62
rørtong[nn], 62
rörtång[sv], 62
røyr[nn], 62
røyrleggjar[nn], 64
råde[nb], 62
råhet[sv], 72
råjern[nb], 61
råmaterial[sv], 69
råmateriale[nb], 69
råolja[sv], 22
råolje[nb], 22
råvara[sv], 69
råvare[nb], 69

S

saaste[fi], 64
sáddobossun[se], 73
sadjin[se], 98
sadjingeađgi[se], 98
sadjit[se], 44
safe-edge file, 36
safety trustee, 73
sag[nb], 73
sagblad[nb], 73
sággeguoskkahat[se], 64
sággejurssan[se], 30
sággi[se], 61
saha[fi], 73
sahá[se], 73
sahádearri[se], 73
sahanterä[fi], 73
sáhcu[se], 74
sáhcunnállu[se], 75
sáhcut[se], 74
sáhpán[se], 56
sáigun[se], 65
sajádat[se], 79
sajáiduhttit[se], 47
sajálduhttit[se], 47
sakka[fi], 42
saks[nb], 74
saksesplint[nb], 83
sakset[fi], 74
salpa[fi], 51
sálvu[se], 41
samanføying[nn], 48
samlestokk[nb], 53
samlingsrör[sv], 53
sammanfogning[sv], 48
sammenføyning[nb], 48
sammutin[fi], 35
sandblasting, 73
sandblästring[sv], 73

sandblåsing[nb], 73
sandpaper, 30
sandpapir[nb], 30
sandpapper[sv], 30
sanka[fi], 9
sarana[fi], 44
sárggon[se], 64
sárggus[se], 27
sárgunbreahtta[se], 27
sarjakytkentä[fi], 76
sarjaliitäntä[fi], 76
sáttobábir[se], 30
saturate, 73
save, 73
saw, 73
saw blade, 73
sax[sv], 74
saxpinne[sv], 83
scale, 73
scanner, 73
scanner[sv], 73
scarf, 68
schablon[sv], 40
scissors, 74
scoop, 8
scouring machine, 74
scrap iron, 74
scrape, 74
scraper, 74
scratch, 74
scratch awl, 75
scratching, 74
screen, 74
screw, 74
screw clamp, 15
screw die, 23
screw extractor, 74
screw feeder, 33
screw pitch, 74
screw pitch gauge, 89
screw tap, 88
screwdriver, 74
scribe, 74
scriber, 75
scribing awl, 75
scroll, 75
seaguhangámmár[se], 55
seaguhanhátna[se], 55
seaguhit[se], 3
seaguhus[se], 3
seaguhusgorri[se], 67
seal, 75
sealing device, 75
sealing face, 19
sealing ring, 75
seam, 37
šearbma[se], 74
seat, 75
seaván[se], 22
seavdinbumpa[se], 33
second tap, 75
secondary, 75
secondary edge, 75
section, 22
securing plate, 75
seegerring[nb], 51
seghet[sv], 92
seghärdat stål[sv], 89
seghärdning[sv], 89
segjärn[sv], 57
segohus[se], 3
seigherda stål[nb], 89
seigherding[nb], 89
seighet[nb], 92
seigjern[nb], 57
seigleik[nn], 92
sekoitin[fi], 55
sekoituskammio[fi], 55
sekskantnøkkel[nb], 44
sekskantskrue[nb], 44
sekundára[se], 75
sekundær[nb], 75
sekundär[sv], 75
self-centering, 75
self-locking screw, 76
self-tapping screw, 76

šelget[se], 64
selvstarter[nb], 84
semi-circular protractor, 76
semiconductor, 76
semi-products, 42
senkhode[nb], 21
senkhovud[nn], 21
sensor, 24
sensor[nb], 24
sensor[sv], 24
sensori[fi], 24
senterbor[nb], 13
senterhøgd[nn], 44
senterhøyde[nb], 44
senterspiss[nb], 88
sentrumsbor[nb], 13
sentrumsvinkel[nb], 13
seos[fi], 3
seossuhde[fi], 67
seostaa[fi], 3
separator, 47
seppä[fi], 80
serial connection, 76
seriekopling[nb], 76
seriekoppling[sv], 76
seriel interface, 76
seriellt gränssnitt[sv], 76
serielt grensesnitt[nb], 76
series connection, 76
serrat[nb], 76
serrat[se], 76
serrate, 76
serrate tool, 76
serratere[nb], 76
serrateret[se], 76
service, 52
service[nb], 52
service[sv], 52
set, 76
set screw, 2, 76
sete[nb], 75
settherding[nb], 12
setting angle, 77
settskrue[nb], 76
sewer pipe, 26
sexkantnippel[sv], 44
sexkantnyckel[sv], 44
sexkantskruv[sv], 44
sfearalaš láger[se], 82
sfearalaš[se], 82
sfærisk lager[nb], 82
sfærisk[nb], 82
sfärisk[sv], 82
sfäriskt lager[sv], 82
shaft, 42, 77
shaft extension, 77
shaft journal, 77
shale, 77
shaper, 63
shaping machine, 63
sharpen, 41
sharpening, 41
shavings, 93
shears, 74
sheath knife, 77
sheet, 63
sheet metal, 77
sheet metal shears, 77
shell end mill, 77
shield gas, 77
shift, 77
shift gear, 13
shim, 78
shock absorber, 78
short-circuit, 78
shoulder, 78
shrink, 78
shrink on, 78
shrinkage, 78
shrinking, 78
shunt, 78
shunt switch, 78
shuntvewntil[sv], 11
sidavbitare[sv], 78
side cutting pliers, 78
side of thread, 78

side[fi], 71
sideaine[fi], 8
sideavbitartong[nn], 78
sideavbitertang[nb], 78
sideprodukt[nb], 11
sidosaine[fi], 8
sieve, 85
signaali[fi], 78
signal, 78
signal horn, 45
signal[nb], 78
signal[sv], 78
signála[se], 78
signalhorn[nb], 45
signalhorn[sv], 45
sihkkarasti[se], 39
sihkkarastinriekkis[se], 80
siipi[fi], 8
siipimutteri[fi], 98
siirto[fi], 92
siirtopylkän holkki[fi], 88
siirtymä[fi], 92
siivilä[fi], 85
sijainti[fi], 79
sikring[nb], 39
sikringsring[nb], 80
sil[nb], 85
sil[sv], 85
silencer, 78, 79
silli[se], 65, 85
silmukka-avain[fi], 81
silmukkaruuvi[fi], 32
siloittaa[fi], 68
silta[fi], 93
siltanosturi[fi], 92
šimir[se], 5
single-acting cylinder, 79
sinkadeapmi[se], 100
sinken[se], 100
sinkilä[fi], 9
sinkitys[fi], 100
sintering, 79
sintraus[fi], 79
sintren[se], 79
sintring[nb], 79
sintring[sv], 79
sirdin[se], 92
sirdinmihtádas[se], 80
sirkelsag[nb], 15
sisasuddadeapmi[se], 98
sisasuddan[se], 98
sisriekkis[se], 46
site, 79
sitkatvuohta[se], 92
sitkeys[fi], 92
situation, 79
sivuleikkurit[fi], 78
sivusärmä[fi], 75
sivutuote[fi], 11
sjablon[nb], 40
självcentrerande[sv], 75
självhämmande skruv[sv], 76
sjølgjengende skrue[nb], 76
sjøllåsende skrue[nb], 76
sjølsentrerende[nb], 75
sjølsperrende skrue[nb], 76
sjølvgjengande skrue[nn], 76
sjølvlåsande skrue[nn], 76
sjølvsentrerande[nn], 75
sjølvstartar[nn], 84
skaft[nb], 42
skaft[sv], 42
skafta[sv], 42
skala[nb], 73
skála[se], 73
skala[sv], 73
skaltång[sv], 86
skanner[nb], 73
skárrit[se], 74
skarvförbindning[sv], 35
skarvsladd[sv], 32
skearrogoazan[se], 26
skearru[se], 26
skenhammare[sv], 79
skid, 80
skieivun[se], 29

skierat[se], 74
-skierat[se], 90
skift[nb], 77
skift[sv], 77
skiftenøkkel[nb], 2
skiftnyckel[sv], 2
skims[nb], 78
skiss[sv], 96
skiva[sv], 26
skivbroms[sv], 26
skivdnjeviŋkil[se], 8
skive[nb], 26
skivebrems[nb], 26
skjefte[nb], 42
skjemt[nb], 8
skjer[nn], 29
skjerebrennar[nn], 23
skjerefart[nn], 23
skjerfil[nn], 42
skjerm[nb], 74
skjutmått[sv], 80
skjutpassning[sv], 79
skjær[nb], 29
skjærebrenner[nb], 23
skjærehastighet[nb], 23
skjærfil[nb], 42
skjøteledning[nb], 32
skoađas[se], 35
skoavhli[se], 65
skohter[se], 80
skootteri[fi], 80
skovel[sv], 8
skovl[nb], 8
skralle[nb], 68
skrapa[sv], 74
skrape[nb], 74
skrapjern[nb], 74
skrivar[nn], 66
skrivare[sv], 66
skriver[nb], 66
skrubbstål[nb], 72
skrubbstål[sv], 72
skrue[nb], 74
skruedrev[nb], 99
skruestikke[nb], 7
skruetransportør[nb], 33
skrueuttrekker[nb], 74
skrueuttrekkjar[nn], 74
skrujern[nb], 74
skrunøkkel[nb], 81
skrutrekker[nb], 74
skrutrekkjar[nn], 74
skrutvinge[nb], 15
skruv[sv], 74
skruvmejsel[sv], 74
skruvnyckel[sv], 81
skruvstycke[sv], 7
skruvtransportör[sv], 33
skruvtving[sv], 15
skruvutdragare[sv], 74
skruvva[se], 74
skruvvabasta[se], 7
skruvvabonjan[se], 74
skruvvačárvvon[se], 15
skruvvadoalan[se], 7
skruvvageasán[se], 74
skruvvaruovdi[se], 74
skruvvenčoavdda[se], 81
skränka[sv], 76
skråavbiter[nb], 78
skuibi[se], 39
skuohppu[se], 10
skyddsgas[sv], 77
skyddsglasögon[sv], 67
skyddshjälm[sv], 67
skyddshylsa[sv], 79
skyddsombud[sv], 73
skyddsutrustning[sv], 67
skyvelære[nb], 80
skyvemål[nb], 80
skyvepasning[nb], 79
skärbrännare[sv], 23
skärdjup[sv], 23
skärhastighet[sv], 23
skärpverktyg[sv], 63
skötsel[sv], 52

skøyteleidning[nn], 32
slag, 79
slag hammer, 79
slagboremaskin[nb], 42
slagborrmaskin[sv], 42
slagg[nb], 79
slagg[sv], 79
slagghammar[nn], 79
slagghammare[sv], 79
slagghammer[nb], 79
slaggpikke[nb], 79
slaglengd[nn], 50
slaglengde[nb], 50
slaglängd[sv], 50
slagseghet[sv], 45
slagseighet[nb], 45
slagseigleik[nn], 45
slakk[nb], 6
šlámbarruovdi[se], 74
slang[sv], 45
slange[nb], 45
slangeklemme[nb], 45
slangklämma[sv], 45
šláŋŋa[se], 45
šláŋŋačárvvon[se], 45
šlápma[se], 57
šleagga[se], 79
šleađggočuovga[se], 36
sledge hammer, 79
sleeve, 10, 79, 80
slegge[nb], 79
sleggje[nn], 79
sleide[nb], 79
slid[sv], 79
slide, 79, 80
slide coupling, 79
slide fit, 79
slide gauge, 80
slide valve, 39
sliding bearing, 80
sliding caliper, 80
sliding friction, 80
sliding resistance, 80
slidkniv[sv], 77
slidventil[sv], 39
slig[nb], 27
slig[sv], 27
šliipa[se], 41
šliipen[se], 41
šliipenbábir[se], 30
šliipengordni[se], 1
šliipenliidni[se], 1
šliipenmálle[se], 41
šliipenmašiidna[se], 41
šliipenmoallu[se], 1
šliipennibba[se], 41
šliipenskearru[se], 41
šliipet[se], 41
šlimpa[se], 60
slip, 80
slipa[sv], 41
slipduk[sv], 1
slipe[nb], 41
slipekorn[nb], 1
slipelerret[nb], 1
slipelære[nb], 41
slipemaskin[nb], 41
slipeskive[nb], 41
slipestein[nb], 41
slipestift[nb], 41
sliping[nb], 41
slipkorn[sv], 1
slipmaskin[sv], 41
slipning[sv], 41
slipping clutch, 79
slipskiva[sv], 41
slipsten[sv], 41
slipstift[sv], 41
slira[sv], 80
slirkoppling[sv], 79
slitage[sv], 96
slitasje[nb], 96
slitestyrke[nb], 96
slitestål[nb], 96
slitsmått[sv], 33
slitstyrka[sv], 96

slitstål[sv], 96
šloaŋkkisteapmi[se], 6
slot, 80
slotted screw, 80
slotting tool, 80
šluppot[se], 53
slure[nb], 80
slurekopling[nb], 79
sluseventil[nb], 39
slussventil[sv], 39
slutbehandling[sv], 34
sluten stömkrets[sv], 16
slutta krins[nn], 16
slutta strømkrets[nb], 16
sluttapp[nb], 64
slägga[sv], 79
släppningsvinkel[sv], 15
släppningsyta[sv], 15
slö[sv], 8
sløv[nb], 8
sløvast[nn], 7
sløves[nb], 7
slåmaskin[nb], 57
slåttermaskin[sv], 57
smáhkku[se], 14
smed[nb], 80
smed[sv], 80
smedja[sv], 80
smelle[nb], 68
smelte[nb], 54
smeltebad[nb], 56
smeltepunkt[nb], 54
smeltetemperatur[nb], 54
smergelduk[sv], 1
smi[nb], 37
smibar[nb], 38
smida[sv], 37
smidbar[sv], 38
smideshärd[sv], 38
smidessvetsning[sv], 38
smidestång[sv], 38
smie[nb], 80
smierusvuohta[se], 10
smiesse[nb], 38
smihammar[nn], 79
smihammer[nb], 79
smilčečiehka[se], 68
smirrodat[se], 10
smisveising[nb], 38
smitang[nb], 38
smith, 80
smithy, 80
smitong[nn], 38
smygvinkel[nb], 8
smygvinkel[sv], 8
smälta[sv], 54
smältpunkt[sv], 54
smälttemperatur[sv], 54
smøre[nb], 51
smørefett[nb], 51
smørefilm[nb], 51
smøremiddel[nb], 51
smørenippel[nb], 41
smøreolje[nb], 52
smøreoljekanne[nb], 58
smørespor[nb], 58
smörja[sv], 51
-smørjar[nn], 58
smørje[nn], 51
smørjefeitt[nn], 51
smørjefilm[nn], 51
smørjemiddel[nn], 51
smørjenippel[nn], 41
smørjeolje[nn], 52
smørjeoljekannen[nb], 58
smørjespor[nn], 58
smörjfett[sv], 51
smörjfilm[sv], 51
smörjmedel[sv], 51
smörjnippel[sv], 41
smörjolja[sv], 52
smörjspår[sv], 58
snabbstål[sv], 44
snáhki[se], 71
snađđen[se], 49
snap ring, 80

sneaktačuohpastat[se], 22
sneaktagovva[se], 22
snekkedrev[nb], 99
snekkehjul[nb], 99
snekker[nb], 12
snekkersag[nb], 48
snekkeskrue[nb], 99
snickare[sv], 12
snickarsåg[sv], 48
snihkkár[se], 12
snihkkársahá[se], 48
snikkar[nn], 12
snikkarsag[nn], 48
snitt[nb], 22
snitt[sv], 22
snow scooter, 80
snäckhjul[sv], 99
snäckskruv[sv], 99
snøggstål[nn], 44
snöskoter[sv], 80
snøskuter[nb], 80
soadjemuhtter[se], 98
soadji[se], 8
soairu[se], 5, 75
socket, 9, 10, 80
socket wrench, 9, 81
sodju[se], 7
soft soldering, 81
šohkka[se], 14
sokka[fi], 61
sokkanaula[fi], 83
sokkel[nb], 38
solar cell, 81
solar[nb], 25
solcell[sv], 81
solcelle[nb], 81
solder, 81
soldering, 81
soldering iron, 81
soldering paste, 37
soldering tin, 81
šolgantemperatuvra[se], 54
šolgenávnnas[se], 37
šolggas[se], 37
šolgu[se], 54
solid, 81
solution, 30
solvent, 81
soot, 81
sorkkarauta[fi], 22, 57
sorvaaja[fi], 93
sorvari[fi], 93
sorvata[fi], 93
sorvi[fi], 49
sorvinterä[fi], 49
sot[nb], 81
sot[sv], 81
sota[sv], 81
sote[nb], 81
sound testing, 81
source of current, 65
sovite[fi], 35
sovituslevy[fi], 78
sŋiran[se], 68
spackel[sv], 82
spackla[sv], 68
spáitastálli[se], 44
spak[nb], 42
spak[sv], 42
span of jaws, 98
spanner, 81
spant[nb], 71
spant[sv], 71
sparaideapmi[se], 95
sparaidit[se], 95
spare part, 82
spark, 82
spark coil, 45
spark plug, 82
spark plug cable, 64
spark plug cap, 82
spark plug wrench, 82
sparkel[nb], 82
sparkel[se], 82
sparkelastit[se], 68
sparkle[nb], 68

spattle, 82
spatula, 82
specifik[sv], 82
speed, 82
speedometer, 82
spel[nn], 98
spel[sv], 6, 98
spelle[se], 77
spelle-[se], 90
-spelle[se], 75
spennbakke[nb], 15
spennblikk[nb], 51
spenne fast[nb], 33
spenne opp[nb], 33
spennhylse[nb], 88
spenning[nb], 89, 96
spennstift[nb], 64
sperre[nb], 51
sperremekanisme[nb], 51
sperring[nb], 95
spesific, 82
spesifikk[nb], 82
spetsvinkel[sv], 1
spett[nb], 22
spett[sv], 22
spherical, 82
spherical bearing, 82
spiikkár[se], 57
spiikkárgeasán[se], 57
spik[sv], 57
spikar[nn], 57
spikartrekkjar[nn], 57
spiker[nb], 57
spikertrekker[nb], 57
spikutdragare[sv], 57
spill[nb], 98
spindel[nb], 82
spindel[sv], 82
spindeldocka[sv], 43
spindeldokke[nb], 43
spindle, 43, 82
spintočuohpan[se], 2
spintu[se], 39
spiral drill, 94
spiral groove, 75
spiralbor[nb], 94
spiralborr[sv], 94
spiralspor[nb], 75
spiralspår[sv], 75
spisstang[nb], 36
spisstapp[nb], 88
spisstong[nn], 36
spissvinkel[nb], 1
spjell[nb], 36
spjäll[sv], 36
spline shaper, 83
splines, 82
splines[nb], 82
splines[sv], 82
splint, 83
splint[nb], 83
splinter, 83
split pin, 83
splittpinne[nb], 83
spole[nb], 16
spole[sv], 16
spon[nb], 14
sponbrytar[nn], 14
sponbryter[nb], 14
sponflate[nb], 6
sponplate[nb], 14
sponskjerande til [nn], 14
sponskjærende bearbeiding[nb], 14
sponta[se], 37
spontalakta[se], 68
spontet[se], 37
sponvinkel[nb], 6
spool, 16, 98
spor[nb], 80
sporkile[nb], 97
sporkulelager[nb], 92
sporskrue[nb], 80
sporstift[nb], 37
spot welder, 83
spot welding, 83

spot welding machine, 83
spray nozzle, 58
sprengskive[nb], 73
spring, 83
spring steel, 83
spring tension, 83
spring washer, 73
sprint[sv], 83
sprödhet[sv], 10
sprøhet[nb], 10
sprøleik[nn], 10
spylespiss[nb], 96
spänna fast[sv], 33
spännback[sv], 15
spänndorn[sv], 15
spännhylsa[sv], 88
spänning[sv], 89, 96
spännklov[sv], 15
spärr[sv], 51
spärranordning[sv], 51
spärrnyckel[sv], 68
spån[sv], 14
spånbrytare[sv], 14
spånplatta[sv], 14
spånskärande bearbetning[sv], 14
spånvinkel[sv], 6
spånyta[sv], 6
spår[sv], 49, 80
spårkullager[sv], 92
spårskruv[sv], 80
square, 83, 84
square head screw, 83
square thread, 36
staattinen[fi], 84
staattori[fi], 84
stabil[nb], 84
stabil[sv], 84
stabilisera[sv], 6
stabilisere[nb], 6
stabilitet[nb], 84
stabilitet[sv], 84
stability, 84
stable, 84
stag[nb], 71
stag[sv], 71
stahtalaš[se], 84
stahtor[se], 84
stainless steel, 84
stake, 4
stállebáddi[se], 98
stálledoalan[se], 91
stálletoavva[se], 98
stálleullu[se], 85
stálli[se], 85
stamp, 67
stamp(er), 67
stámpa[se], 19
stamping machine, 67
stand, 84
stansa[sv], 67
stanse[nb], 67
stansemaskin[nb], 67
stanseverktøy[nb], 67
stansmaskin[sv], 67
stansverktyg[sv], 67
stáđis[se], 84
stáđisvuohta[se], 84
stáđđi[se], 4
star-delta connection, 84
starga[se], 84
stargatvuohta[se], 84, 85
stargodat[se], 85
start, 84
starta[sv], 84
starte (bil[nb], 84
starter, 84
starter gear ring, 84
starting motor, 84
starting point, 56
startkrans[nb], 84
startkrans[sv], 84
startmotor[nb], 84
startmotor[sv], 84
static, 84
statisk[nb], 84

statisk[sv], 84
stativ[nb], 84
stativ[sv], 84
stator, 84
stator[nb], 84
stator[sv], 84
steady, 84
stealládat[se], 38
steallečiehka[se], 84
steallefiilu[se], 36
steallenihppel[se], 44
stealleviŋkil[se], 84
stealli[se], 78, 86
steam, 85
steam engine, 85
steamppal[se], 62
steel, 85
steel shavings, 85
steel tape rule, 54
steel wool, 85
steering, 85
steering wheel, 42
steghjul[sv], 68
steglös[sv], 85
stellenskruvva[se], 2
stem, 82
stempel[nb], 62
stempelring[nb], 62
stempelslag[nb], 86
stempelstang[nb], 62
stepless, 85
sticka av[sv], 22
stickpropp[sv], 64
stickstål[sv], 80
sticksåg[sv], 89
stielkalakta[se], 35
stiffness, 85
stift[nb], 19
stigning[nb], 74
stikke av[nb], 22
stikkontakt[nb], 80
stikkpasser[nb], 26
stikksag[nb], 89
stikkstål[nb], 80
stillbar brotsj[nb], 2
stillbar vinkel[nb], 8
stillingsenergi[nb], 65
stillskrue[nb], 2
stirddisvuohta[se], 85
stirrup, 9
stivhet[nb], 85
stivleik[nn], 85
stivren[se], 85
stivrenbielka[se], 7
stivrenteknihkka[se], 20
stivrran[se], 20
stjernefastnøkkel[nb], 17
stjernenøkkel[nb], 81
stjerneskrujern[nb], 22
stjernetrekantkopling[nb], 84
stjärntriangelkoppling[sv], 84
stock, 42
-stong[nn], 62
stoppskive[nb], 96
stoppskruv[sv], 76
straight pane hammer, 85
straightening, 3
straightening tool, 85
strain, 60
strainer, 34, 85
strap, 27
straum[nn], 22
straum-[nn], 2, 26
straumkjelde[nn], 65
straumkrins[nn], 30
straumretning[nn], 25
streckmått[sv], 53
strekkfasthet[nb], 89
strekkfastleik[nn], 89
strekmål[nb], 53
strength, 86
strengthen, 86
strengthening, 86
stress, 89
stripping pliers, 86
stroke, 86

structural steel, 19
strupar[nn], 14
struper[nb], 14
strupeventil[nb], 90
strutting, 71
strykmått[sv], 53
strypventil[sv], 90
sträng[sv], 7
strøm[nb], 22
ström[sv], 22
strømfordeler[nb], 26
strømkilde[nb], 65
strømkrets[nb], 30
strömkrets[sv], 30
strömkälla[sv], 65
strømretning[nb], 25
strömriktning[sv], 25
stud, 60
stud torque, 86
stuka[sv], 16
stuke[nb], 16
stukesveising[nb], 10
stuksvetsning[sv], 10
stycklista[sv], 60
stykkliste[nb], 60
styreeining[nn], 20
styreenhet[nb], 20
styrejern[nb], 96
styrenhet[sv], 20
styring[nb], 85
styringsteknikk[nb], 20
styrning[sv], 85
styrteknik[sv], 20
styvhet[sv], 85
städ[sv], 4
ställbar brotsch[sv], 2
ställskruv[sv], 2
ställvinkel[sv], 8, 77
stöd[sv], 86
støpe[nb], 12
støpegods[nb], 12
støpejern[nb], 12
støperi[nb], 38
støpsel[nb], 64
störtkylning[sv], 68
støtdemper[nb], 78
stötdämpare[sv], 78
støtfanger[nb], 10
stötfångare[sv], 10
støvsugar[nn], 94
støvsuger[nb], 94
støy[nb], 57
støype[nb], 12
støypegods[nb], 12
støypejer[nn], 12
støyperi[nb], 38
støytdempar[nn], 78
støytfangar[nn], 10
stål[nb], 85
stål[sv], 85
stålfäste[sv], 91
stålhaldar[nn], 91
stålholder[nb], 91
stålhållare[sv], 91
stålull[nb], 85
stålull[sv], 85
suction, 86
suddadit[se], 54
suddat[se], 54
sudjenjealbma[se], 67
sudjenláset[se], 67
sudjenneavvut[se], 67
sug[nb], 86
sug[sv], 86
sugadangiehta[se], 95
sula[fi], 54
sulaa[fi], 54
sulake[fi], 39
sulamispiste[fi], 54
sulattaa[fi], 54
sulatuserä[fi], 54
suljettu piiri[fi], 16
sulkutulppa[fi], 60
sulphur, 86
sumuttuminen[fi], 5
sumutus[fi], 5

suodatin[fi], 34, 85
suodattaa[fi], 60
suodatus[fi], 34
suodjalusáittardeaddji[se], 73
suodjegássa[se], 77
suoidnespábbačárvvon[se], 43
suoja-asu[fi], 67
suojaholkki[fi], 79
suojakaasu[fi], 77
suojakypärä[fi], 67
suojalasit[fi], 67
suojapakat[fi], 51
suojavarusteet[fi], 67
šuokŋaiskan[se], 81
suonjarbovrenmašiidna[se], 68
suonjarháltásaš[se], 68
suonjarláger[se], 68
suođđu[se], 50
suorakulma[fi], 83
suorakulmain[fi], 84
suorgebohcci[se], 53
suoritin[fi], 66
suorresággi[se], 83
supistusholkki[fi], 88
suppilo[fi], 39
support, 86
surface, 86
surface cutter, 32
surface gauge, 86
surface grinding machine, 86
surface norm, 87
surface of fracture, 87
surface plate, 87
surface roughness, 87
surface rule, 87
surface treatment, 87
surfacing, 16
suunta[fi], 25
suuntaispala[fi], 40
suuntapiirrin[fi], 53, 86
suuntaventtiili[fi], 25
suunturi[fi], 53
suutin[fi], 57, 58, 96
suutinventtiili[fi], 96
suvrenanu stálli[se], 1
svarv[sv], 49
svarva[sv], 93
svarvare[sv], 93
svarvstål[sv], 49
svavel[sv], 86
sveis[nb], 97
sveisa[se], 97
sveisar[nn], 97
sveisašolgu[se], 56
sveise[nb], 97
sveiseapparat[nb], 97
sveisebend[nb], 11
sveiseblindhet[nb], 4
sveiseblink[nb], 4
sveisebrennar[nn], 97
sveisebrenner[nb], 97
sveiseelektrode[nb], 97
sveisefuge[nb], 97
sveisehjelm[nb], 97
sveisejeaddji[se], 97
sveiselarve[nb], 7
sveisemaske[nb], 97
sveisen[se], 97
sveisenapparáhta[se], 97
sveisenárpu[se], 98
sveisenboalddan[se], 97
sveisenelektroda[se], 97
sveisengurra[se], 41, 97
sveisenhápma[se], 97
sveisenluodda[se], 97
sveisenmáska[se], 97
sveisenmoalki[se], 11
sveisenrievdadeaddji[se], 97
sveisenšleađggastat[se], 4
sveisenstreaŋga[se], 7
sveisensuddon[se], 4
sveisensuotna[se], 7
sveisentransformáhtor[se], 97
sveiseomformer[nb], 97
sveiser[nb], 97

sveisestreng[nb], 7
sveiset[se], 97
sveisetråd[nb], 98
sveising[nb], 97
sveiv[nb], 21
sverd[nb], 41
svets[sv], 97
svetsa[sv], 97
svetsaggregat[sv], 97
svetsare[sv], 97
svetsblänk[sv], 4
svetsbrännare[sv], 97
svetselektrod[sv], 97
svetsfog[sv], 97
svetshjälm[sv], 97
svetsinsats[sv], 11
svetsmask[sv], 97
svetsning[sv], 97
svetsomformare[sv], 97
svetssmälta[sv], 56
svetstråd[sv], 98
svinghjul[nb], 37
svingjern[nb], 25
svovel[nb], 86
svänghjul[sv], 37
svängjärn[sv], 25
svärd[sv], 41
swarf, 87
swift, 69
switch, 87
swivel handle, 87
swivel rod, 87
swivel socket wrench, 87
sydän[fi], 98
sydänvoitelu[fi], 98
syke[fi], 67
syklus[nb], 23
syl[nb], 5
syl[sv], 5
sylinddar[se], 23
sylinddargorut[se], 23
sylinddarsuođđaniskkan[se], 24
sylinder[nb], 23
sylinderblokk[nb], 23
sylinderlekkasjetester[nb], 24
sylinteri[fi], 23
sylinterinkansi[fi], 23
sylinteriryhmä[fi], 23
symmetralaš[se], 87
symmetric, 87
symmetrinen[fi], 87
symmetrisk[nb], 87
symmetrisk[sv], 87
synchronous, 87
synchronous motor, 87
synkron[nb], 87
synkron[sv], 87
synkronimoottori[fi], 87
synkroninen[fi], 87
synkronmohtor[se], 87
synkronmotor[nb], 87
synkronmotor[sv], 87
synkruvdnalaš[se], 87
synteettinen[fi], 87
syntetalaš[se], 87
syntetisk[nb], 87
syntetisk[sv], 87
synthetic, 87
syrafast stål[sv], 1
syrefast stål[nb], 1
systeemi[fi], 88
system, 88
system[nb], 88
system[sv], 88
systema[se], 88
syttyvä[fi], 17
sytytustulpan johto[fi], 64
sytytyslaitteisto[fi], 45
sytytyspuola[fi], 45
sytytystulpan hattu[fi], 82
sytytystulppa[fi], 82
syöpyminen[fi], 20
syöttää[fi], 33
syöttö[fi], 33
syöttöakseli[fi], 33

syöttöpumppu[fi], 33
syöttöpyörä[fi], 68
syöttöruuvi[fi], 33
syövyttää[fi], 31
sähkö[fi], 30
sähköasentaja[fi], 30
sähkögeneraattori[fi], 2
sähköinen[fi], 29, 30
sähköverkko[fi], 52
säiliö[fi], 19
säkring[sv], 39
säkringsring[sv], 80
säppi[fi], 51
säppipyörä[fi], 68
särmäyskone[fi], 37
säte[sv], 75
säteis-[fi], 68
säteislaakeri[fi], 68
säteisporakone[fi], 68
säteittäinen[fi], 68
sätthärdning[sv], 12
säädettävä kalvain[fi], 2
säädin[fi], 70
säätäjä[fi], 70
säätö[fi], 40
säätökulmain[fi], 8
säätöpelti[fi], 36
säätöruuvi[fi], 2
säätötekniikka[fi], 19
søkjar[nn], 33
søppeldunk[nb], 39
søyleboremaskin[nb], 94
såg[sv], 73
sågblad[sv], 73

T

T- handle, 89
taajuus[fi], 38
table, 73
tachometer, 71
tacka[sv], 46
tackjärn[sv], 61
tackle, 88
tahdistamaton[fi], 5
tahdistettu[fi], 87
tahko[fi], 41
tahti[fi], 86
tahtimoottori[fi], 87
tail stock, 88
tail stock centre, 88
tail stock spindle, 88
taittaa[fi], 37
taivuttaa[fi], 37
taivutuskone[fi], 37
taka-akseli[fi], 69
takaiskuventtiili[fi], 13
takapyörä[fi], 69
takatuli[fi], 48
takoa[fi], 37
takt[nb], 86
takt[sv], 86
-tal[nn], 72
talja[fi], 88
talja[sv], 88
talje[nb], 88
tallentaa[fi], 73
taltata[fi], 14
taltta[fi], 14
tang, 88
tang[nb], 91
tange[nb], 88
tangentbord[sv], 48
tank, 19
tank[nb], 19
tank[sv], 19
tánka[se], 19
tanndeling[nb], 15
tannhjul[nb], 40
tannhøgd[nn], 91
tannhøyde[nb], 91
tannstang[nb], 68
tannstong[nn], 68
tanntal[nn], 58
tanntall[nb], 58
taottava[fi], 38
tap, 16, 88

tapahtumasarja[fi], 66
taper, 88
taper clamping sleeve, 88
taper pin, 88
taper tap, 88
taper turning, 88
tapered, 18
tapered sleeve, 88
tapp[nb], 41
tapp[sv], 41
tappi[fi], 41, 61
tappijyrsin[fi], 30
tarkastaa[fi], 19
tarkastus[fi], 46
tarkistaa[fi], 2
tarkistus[fi], 2
tarkkuus[fi], 65
T-arm[nb], 89
tasainen[fi], 36
tasapaino[fi], 6
tasapainoinen[fi], 84
tasapainottaa[fi], 6
tasasuuntaaja[fi], 69
tasata[fi], 17
tasauspyörastön lukitus[fi], 25
tasauspyörästö[fi], 25
tasavirta[fi], 25
taso[fi], 36
tasohiomakone[fi], 86
tasoitta[fi], 63
tasojyrsin[fi], 32
tasojyrsinkone[fi], 45
tasolaikka[fi], 32
tasosorvata[fi], 32
tastatur[nb], 48
tegnebrett[nb], 27
tegning[nb], 27
tehdas[fi], 32
teho[fi], 29
tehonsiirtohihna[fi], 92
teiknebrett[nn], 27
teikning[nn], 27
tela[fi], 7
telamatto[fi], 7
telescopic gauge, 89
teleskohpamihtádas[se], 89
teleskopmål[nb], 89
teleskopstickmått[sv], 89
-teljar[nn], 71
temper, 43
tempered steel, 89
tempering, 43, 89
tempering furnace, 42
template, 40
tenn[sv], 90
tenning[nb], 45
tenningsanlegg[nb], 45
tenningsboks[nb], 13
tenningssystem[nb], 45
tennlödning[sv], 81
tennplugg[nb], 82
tennplugg-[nb], 64
tennpluggnøkkel[nb], 82
tennspole[nb], 45
tenon saw, 89
tensile strength, 89
tension, 89
teroittaa[fi], 44
teroittaminen[fi], 98
teroituslaite[fi], 63
ters[nb], 67
terä[fi], 29
teräkulma tulkki[fi], 41
teräkulma[fi], 22
terälaippa[fi], 41
terän kuoppakuluminen[fi], 62
teränpidin[fi], 91
teräs[fi], 85
teräsvilla[fi], 85
test, 89
test[nb], 89
-testar[nn], 24
tetning[nb], 75
tetningsflate[nb], 19
tetningsring[nb], 75
tetthet[nb], 24

tetting[nb], 75
tettingsflate[nb], 19
tettingsring[nb], 75
tettleik[nn], 24
T-handtag[sv], 89
T-handtak[nb], 89
thinner, 89
thread, 89
thread gauge, 89
thread tap, 88
threading die, 23
threading tool, 13
three phase motor, 90
throttle valve, 90
thrust, 56
thrust bearing, 90
thyristor, 90
tiedosto[fi], 34
tieto[fi], 24
tietokone[fi], 18
tight fit, 90
tighten, 33
tiheys[fi], 24
tiiviste leikkuri[fi], 2
tiiviste[fi], 39, 75
tiivisterengas[fi], 75
tiivistin[fi], 75
tiivistyspinta[fi], 19
tilarbeide[nn], 65
tilarbeiding[nn], 35
tilbakeslag[nb], 48
tilbakeslagsventil[nb], 13
tillsatsmaterial[sv], 34
tilsatsmateriale[nb], 34
tilsett[nb], 34
tilsettmateriale[nb], 34
tiltrekkingsmoment[nb], 86
timing belt, 69
timing chain, 69
timmerman[sv], 48
tin, 90
tin plate, 77
tin shears, 90
tin soldering, 81
tina[fi], 90
tinajuotos[fi], 81
tinman, 90
tinn[nb], 90
tinnlodding[nb], 81
tinsmith, 90
tinsnips, 90
title block, 90
tittelfelt[nb], 90
T-nađđa[se], 89
toajakierre[fi], 34
toisio-[fi], 75
tokomponentlim[nb], 94
tolerance, 90
tolerance of fit, 35
tolerance range, 91
tolerans[sv], 90
toleránsa[se], 90
toleránsaguovlu[se], 91
toleranse[nb], 90
toleranseområde[nb], 91
toleransområde[sv], 91
toleranssi[fi], 90
toleranssialue[fi], 91
tolk[nb], 39
tolk[sv], 39
tollekniv[nb], 77
tomgang[nb], 45
tomgangsspenning[nb], 58
tomgång[sv], 45
tomgångsspänning[sv], 58
tommegjenge[nb], 46
tommestokk[nb], 46
tong[nn], 91
-tong[nn], 17, 71
tongs, 91
tool, 91
tool box, 91
tool cutting edge inclination, 68
tool holder, 91
tool path, 91

toolmaker, 91
tooth depth, 91
tooth pitch, 15
toothed bar, 68
toothed wheel, 40
topplock[sv], 23
topplokk[nb], 23
torada lager[nb], 28
torque, 91
torque wrench, 91
torvi[fi], 45
totaktsmotor[nb], 94
toughness, 92
track, 7
track roller bearing, 92
tractor, 92
trakt[nb], 39
traktor[nb], 92
traktor[se], 92
traktor[sv], 92
traktordáigu[se], 43
traktori[fi], 92
transformáhtor[se], 92
transformator[nb], 92
-transformator[nb], 97
transformator[sv], 92
transformer, 92
transistor, 92
transistor[nb], 92
transistor[se], 92
transistor[sv], 92
transistori[fi], 92
transition fit, 92
transmisjon[nb], 92
transmission, 40, 92
transmission belt, 92
transmission[sv], 92
transmišuvdna[se], 92
transom, 93
transporter, 20
transportør[nb], 20
transportör[sv], 20
trapesajeaŋga[se], 92
trapesgjenge[nb], 92
trapetsgänga[sv], 92
trapetsikierre[fi], 92
trapezoid thread, 92
trash barrel, 39
tratt[sv], 39
travelling trolley, 92
travers[nb], 93
travers[sv], 93
traverse, 93
traverse crane, 92
traverskran[nb], 92
traverskran[sv], 92
traversvagn[sv], 92
traversvogn[nb], 92
treat, 65
trefasemotor[nb], 90
trefasmotor[sv], 90
trekantfil[nb], 93
trekantsfil[sv], 93
trekkbrosj[nb], 83
trekt[nb], 39
treskrue[nb], 99
triangular file, 93
trim, 93
trimma[sv], 93
trimme[nb], 93
trimming machine, 37
trinnlaus[nb], 85
trinnløs[nb], 85
trinse[nb], 72
trissa[sv], 72
trommel[nb], 28
trommelbrems[nb], 28
trukkilava[fi], 70
trumbroms[sv], 28
trumma[sv], 28
trunnion, 41
try square, 84
tryckknapp[sv], 67
trycklager[sv], 90
tryckluft[sv], 17
tryckluftshammare[sv], 3

tryckreduceringsventil[sv], 66, 69
tryckregulator[sv], 66
trykkbegrensingsventil[nb], 66
trykkluft[nb], 17
trykkmålar[nn], 66
trykkmåler[nb], 66
trykknapp[nb], 67
trykkregulator[nb], 66
träskruv[sv], 99
tråddragning[sv], 99
trådtrekking[nb], 99
tuki[fi], 71, 86
tulkki[fi], 39, 40
tulostin[fi], 66
tulpanjohto[fi], 64
tulppa[fi], 60
tulppa-avain[fi], 82
tulppaliitin[fi], 64
tumgänga[sv], 46
tumstock[sv], 46
tunkeuma[fi], 98
tunkki[fi], 47
tunnussana[fi], 60
turbiidna[se], 93
turbiini[fi], 93
turbin[nb], 93
turbin[sv], 93
turbine, 93
turbo charger, 93
turbo compressor, 93
turbo[fi], 93
turbo[nb], 93
turbo[se], 93
turbo[sv], 93
turbokompressor[nb], 93
turbulence, 93
turbulens[nb], 93
turbulens[sv], 93
turbulensi[fi], 93
turn, 93
turn (of a winding), 93
turner, 93
turning moment, 91
turning tool, 49
turnings, 93
turtall[nb], 72
turteller[nb], 71
tuuletin[fi], 33
tuulettimen hihna[fi], 33
tuumakierre[fi], 46
tuurna[fi], 27
tverregg[nb], 49
tverrsleide[nb], 22
tverrsnitt[nb], 22
tversavbitar[nn], 57
tversavbiter[nb], 57
tvärbalk[sv], 93
tväregg[sv], 49
tvärslid[sv], 22
T-väännin[fi], 89
tvåkomponentlim[sv], 94
tvåradigt lager[sv], 28
tvåtaktsmotor[sv], 94
tweezers, 62
twist, 98
twist drill, 94
twisting moment, 91
two component glue, 94
two-stroke engine, 94
tyhjennystulppa[fi], 27
tyhjäkäyntijännite[fi], 58
tylsyä[fi], 7
tylsä[fi], 8
tynnar[nn], 89
tynner[nb], 89
tyre, 94
tyristor[nb], 90
tyristor[se], 90
tyristor[sv], 90
tyristori[fi], 90
tyssähitsaus[fi], 10
tyssätä[fi], 16
työkalu[fi], 91
työkalulaatikko[fi], 91

työkaluntekijä[fi], 91
työkalupakki[fi], 91
työkappale[fi], 99
työkone[fi], 19, 99
työmaa[fi], 63
työntömitta[fi], 80
työntötiukkuus[fi], 79
työntötuurna[fi], 29
työpaja[fi], 80
työstää[fi], 65
työstö[fi], 35
työsuojeluasiamies[fi], 73
työtahti[fi], 65
työvara[fi], 99
työympäristö[fi], 99
täckplatta[sv], 11
tähtikolmiokytkentä[fi], 84
täljkniv[sv], 77
tändspole[sv], 45
tändstift[sv], 82
tändstiftskabel[sv], 64
tändstiftsnyckel[sv], 82
tändsystem[sv], 45
tärinä[fi], 95
täristä[fi], 95
täryttää[fi], 95
täsmällisyys[fi], 65
täthetsmätare[sv], 24
tätning[sv], 75
tätningsbricka[sv], 75
tätningsyta[sv], 19
täyteaine[fi], 34
täyteinen[fi], 81
täytelevy[fi], 78
tømrar[nn], 48
tømrer[nb], 48
tång[sv], 91
tånge[sv], 88

U

ubalanse[nb], 94
uimuri[fi], 36
ukvass[nb], 8
ulkonema[fi], 78
ulostyönnin[fi], 29
ulosvedin[fi], 67
unbalance, 94
unbrakonøkkel[nb], 44
underhåll[sv], 52
undertryck[sv], 51
undertrykk[nb], 51
union, 94
union[nb], 94
unit of measure, 94
uniuvdna[se], 94
universal joint, 94
universal milling machine, 94
universal setting gauge, 94
universalfresemaskin[nb], 94
universalfräsmaskin[sv], 94
universalknut[sv], 94
universalledd[nb], 94
universaltang[nb], 17
U-nyckel[sv], 59
up milling, 20
upottaa[fi], 21
upotuspora[fi], 21
upphetta[sv], 43
uppokanta[fi], 21
upprymma[sv], 69
upprymmare[sv], 69
uppvärma[sv], 43
upright drilling machine, 94
upset, 16
ura[fi], 49, 80
urakuulalaakeri[fi], 92
uraruuvi[fi], 80
uratappi[fi], 37
u-ringnyckel[sv], 17
urittaa[fi], 76
uritus[fi], 49
uta(n)bords-[nn], 59
utbalancera[sv], 6
utboringshode[nb], 9
utboringshovud[nn], 9
ut-eining[nn], 59

utenbordsmotor[nb], 59
ut-enhet[nb], 59
utenhet[sv], 59
utjämna[sv], 17
utkraging[nb], 35
utlopp[sv], 59
utmatting[nb], 33
utmattning[sv], 33
utombordsmotor[sv], 59
utslagar[nn], 29
utslager[nb], 29
utsprång[sv], 71
utstötare[sv], 29
utveksling[nb], 40
utvendig nøkkel[nb], 44
utväxling[sv], 40
uudelleen ladattava[fi], 69
uurre[fi], 37

V

vaahdotus[fi], 36
vaakasuora[fi], 44
vaarna[fi], 61
vacuum cleaner, 94
vadjanmašiidna[se], 67
vadjat[se], 67
vadnil[se], 29
váfistansoađis[se], 42
vagn[sv], 96
váhttar[se], 50
vahvike[fi], 5
vahvistaa[fi], 86
vahvistaminen[fi], 86
vahvistin[fi], 3
vahvistus[fi], 86
vahvuus[fi], 86
váibadahttin[se], 33
váidudeapmi[se], 22
váidudit[se], 22
vaier[nb], 98
vaihde[fi], 40
vaihdettava kovametalliterä-pala[fi], 46
vaihe[fi], 61
vaihesiirto[fi], 61
vaihtaa[fi], 13
vaihteisto[fi], 40
vaihtovirta[fi], 3
vaijeri[fi], 98
váikkuhusmearri[se], 29
vaimennin[fi], 22
vaimentaa[fi], 22
vajan[se], 67
vajer[sv], 98
vakaa[fi], 84
vakavuus[fi], 84
valaa[fi], 12
valaistusvoimakkuus[fi], 45
valaisu[fi], 45
valanne[fi], 12
váldoávju[se], 52
valikko[fi], 54
valimo[fi], 38
válla[se], 57
valljudanmálle[se], 48
válloruopma[se], 33
vállu[se], 33
valmistaa[fi], 65
valokaari[fi], 29
vals[nb], 24
vals[sv], 24
válsa[se], 24
válsadoaimmahat[se], 72
válsajurssan[se], 24
válsamašiidna[se], 72
valse[nb], 24, 72
valsemaskin[nb], 72
valseverk[nb], 72
valsfres[nb], 24
valsfräs[sv], 24
valsmaskin[sv], 72
valssaamo[fi], 72
valssain[fi], 72
valssi[fi], 24
valssikone[fi], 72
valssilaitos[fi], 72

valsverk[sv], 72
valukappale[fi], 12
valurauta[fi], 12
valve, 94
valve disc, 36
valve lever, 95
valve rocker, 95
valve seat, 95
valve tappet, 95
vane, 8
vange[nb], 7
vann(av)kjølt[nb], 96
vanne[fi], 71
vannesaha[fi], 6
vannpumpetang[nb], 96
vannrett[nb], 44
vannutskiller[nb], 96
varaosa[fi], 82
varata[fi], 13
variaattori[fi], 95
variáhtor[se], 95
variator, 95
variator[nb], 95
variator[sv], 95
varme opp[nb], 43
varmebehandling[nb], 43
varmebestandig[nb], 43
varmefast[nb], 43
varmeverdi[nb], 43
varmistuslevy[fi], 75
varoke[fi], 39
varsellys[nb], 36
varsi[fi], 42
varsijyrsin[fi], 30
varttaa[fi], 42
varuste[fi], 35
varv[sv], 70
várve[se], 49
várvenbeaŋka[se], 49
várvenstálli[se], 49
várvet[se], 93
varvmätare[sv], 71
varvtal[sv], 72
várvvár[se], 93
vasara[fi], 42
vass(av)kjølt[nn], 96
vasspumpetong[nn], 96
vassrett[nn], 44
vastahitsaus[fi], 71
vastajyrsintä[fi], 20
vastamutteri[fi], 21
vastapaino[fi], 6
vastaventtiilli[fi], 13
vastavoima[fi], 70
vastus[fi], 70
vastustuskyky[fi], 70
vater[nb], 50
vaterpass[nb], 50
vátnanbeaŋka[se], 49
vátnat[se], 93
vatnutskiljar[nn], 96
vattenavskiljare[sv], 96
vattenkyld[sv], 96
vattenpass[sv], 50
vattenpumptång[sv], 96
vauhti[fi], 82
vauhtipyörä[fi], 37
vaunu[fi], 96
vávdna[se], 96
vávlet[se], 9
vávlu[se], 9
V-belt, 18
veadjoárja[se], 65
veaika[se], 98
veaikavuoidan[se], 98
veaiki[se], 20
veaiva[se], 21
veaktastággu[se], 50
veallut[se], 44
vealta[se], 64
vealujursanmašiidna[se], 45
veažir[se], 42
vecka[sv], 37
vedenerotin[fi], 96
vedlikehald[nn], 52
vedlikehold[nb], 52

vedsag[nb], 99
vehicle, 95
veike[nb], 98
veikesmørjing[nn], 98
veitsi[fi], 49
veitsiterä[fi], 49
veiv[nb], 21
veivaksel[nb], 21
veive[se], 21
veiveáksil[se], 21
veiveculci[se], 21
veivekássa[se], 21
veivestággu[se], 62
veiveviessu[se], 21
veivhus[nb], 21
veivi[fi], 21
veivkasse[nb], 21
veivstang[nb], 62
veivtapp[nb], 21
veke[nb], 98
veke[sv], 98
vekesmøring[nb], 98
vekselstraum[nn], 3
vekselstrøm[nb], 3
veksle[nb], 13
veksmörjning[sv], 98
vektstang[nb], 50
vektstong[nn], 50
velocity, 82
vendeskjer[nn], 46
vendeskjær[nb], 46
vengemutter[nn], 98
venstre-[nb], 50
vent pipe, 37
ventiila[se], 94
ventiilačohkkehat[se], 95
ventiilalovttan[se], 95
ventil[nb], 94
ventil[sv], 94
ventilasjon[nb], 95
ventilation, 95
ventilation[sv], 95
ventillyftare[sv], 95
ventilløftar[nn], 95
ventilløfter[nb], 95
ventilsete[nb], 95
ventilsäte[sv], 95
venting, 95
venttiili[fi], 94
venttiilin istukka[fi], 95
venttiilinnostin[fi], 95
venttiilinvipu[fi], 95
venytys[fi], 35
verkkojännite[fi], 52
verkkovirta[fi], 52
-verknad[nn], 11
verknadsgrad[nn], 29
verkningsgrad[sv], 29
verkstad[nn], 99
verkstad[sv], 99
verksted[nb], 99
verkty[nn], 91
verktyg[sv], 91
verktygshållare[sv], 91
verktygslåda[sv], 91
verktygsmakare[sv], 91
verktygsväg[sv], 91
verktøy[nb], 91
verktøybane[nb], 91
verktøyfeste[nb], 91
verktøykasse[nb], 91
verktøykiste[nb], 91
verktøymakar[nn], 91
verktøymaker[nb], 91
vernebriller[nb], 67
vernehjelm[nb], 67
verneombod[nn], 73
verneombud[nb], 73
verneutstyr[nb], 67
vernier, 95
vernier caliper, 80
vertical, 61
vertical drilling machine, 94
vertikal[nb], 61
vertikal[sv], 61
vesijäähdytteinen[fi], 96

vesipumppupihdit[fi], 96
vesivaaka[fi], 50
vetolujuus[fi], 89
vetopyörä[fi], 28
vev[sv], 21
vevaxel[sv], 21
vevhus[sv], 21
vevstake[sv], 62
vevtapp[sv], 21
vian paikannus[fi], 33
vibrasjon[nb], 95
vibrate, 95
vibration, 95
vibration[sv], 95
vibrera[sv], 95
vibrere[nb], 95
vice, 7
viemäri[fi], 59
viemäriputki[fi], 26
vierintälaakeri[fi], 72
vierintäympyrä[fi], 62
view, 96
vifte[nb], 33
viftereim[nb], 33
vihkeohcan[se], 33
vihkket[se], 76
viila[fi], 34
viilaharja[fi], 34
viilain[fi], 34
viilata[fi], 34
viimeistely[fi], 34
viimeistelytappi[fi], 64
viisár[se], 46
viiste[fi], 8
viistää[fi], 8
viivain[fi], 72
viivoitin[fi], 72
vikke[nb], 76
vikle[nb], 98
vikling[nb], 98
vikša[se], 69
vilkkuvalo[fi], 36, 47
vinding[nb], 93
vinge[sv], 8
vingemutter[nb], 98
vingmutter[sv], 98
vinjubasttat[se], 78
vinkel[nb], 3, 83
vinkel[sv], 3, 83
vinkelhake[nb], 84
vinkelhake[sv], 84
vinkeljern[nb], 4
vinkeljärn[sv], 4
vinkelskrujern[nb], 4
vinkelskruvmejsel[sv], 4
vinkelslipar[nn], 4
vinkelsliper[nb], 4
vinkelslipmaskin[sv], 4
vinkeltransportør[nb], 76
vinkeltransportör[sv], 76
vinkeltrekker[nb], 4
vinkeltrekkjar[nn], 4
vinsch[sv], 98
vinsj[nb], 98
vinta[se], 21, 98
vinte[se], 98
vintturi[fi], 69, 98
viđjedoaibma[se], 13
viđjejuvla[se], 13
viđjelássa[se], 13
viđjelohkka[se], 13
viđji[se], 13
viŋkil[se], 83
viŋkilšliipa[se], 4
vipparm[sv], 95
vippe[nb], 50
vippearm[nb], 95
vipu[fi], 42, 50
vipuvarsi[fi], 50
virittää[fi], 93
virkningsgrad[nb], 29
virranjakaja[fi], 26
virransuunta[fi], 25
virta[fi], 22
virtalähde[fi], 65
virtapiiri[fi], 30

virtausmittari[fi], 36
virtavastusventtiili[fi], 90
virvelbildning[sv], 93
visar[nn], 46
visare[sv], 46
viscosity, 96
viser[nb], 46
viskositeetti[fi], 96
viskositehta[se], 96
viskositet[nb], 96
viskositet[sv], 96
vogn[nb], 96
void, 65
voidella[fi], 51
voidesumutin[fi], 58
voimaleikkurit[fi], 9
voiman ulosotto[fi], 65
voimanottolaite[fi], 65
voimansiirto[fi], 65
voiteluaine[fi], 51
voiteluainekerros[fi], 51
voitelukannu[fi], 58
voitelunippa[fi], 41
voitelupuristin[fi], 41
voitelurasva[fi], 51
voiteluruisku[fi], 41
voitelu-ura[fi], 58
voiteluöljy[fi], 52
voltage, 96
vridmoment[sv], 91
vrimoment[nb], 91
vuloštus[se], 38
vuodjinfievru[se], 95
vuogádat[se], 88
vuohču[se], 50
vuoidangurra[se], 58
vuoidanluodda[se], 58
vuoidannihppel[se], 41
vuoidat[se], 51
vuoiddanas[se], 51
vuoiddas[se], 51
vuoiddasassi[se], 51
vuoiddasbožán[se], 41
vuoiddasfilbma[se], 51
vuoiddasolju[se], 52
vuoktabohccedoaibma[se], 11
vuoktasoađis[se], 42
vuolahas[se], 14
vuolahasčiehka[se], 6
vuolahasčuohppan[se], 14
vuolahasdoaján[se], 14
vuolahasduolbadas[se], 14
vuolahasolggoš[se], 6
vuolahaspláhtta[se], 14
vuolahastin[se], 14
vuolggahangierdu[se], 84
vuolggahanmohtor[se], 84
vuolggahit[se], 84
vuolledeatta[se], 51
vuollegisdeatta[se], 51
vuođđo-[se], 66
vuođđobiire[se], 66
vuođđogealdu[se], 30
vuođđomihttu[se], 56
vuorká[se], 34
vuoro[fi], 77
vuorru[se], 77
vuossu[se], 7
vuostedeaddu[se], 6
vuostejursan[se], 20
vuostemuhtter[se], 21
vuostenákca[se], 70
vuostesveisen[se], 71
vuosttaldus[se], 70
vuoto[fi], 50
vuotomittari[fi], 24
vurket[se], 73
väggkontakt[sv], 80
vägguttag[sv], 80
väkipyörä[fi], 8, 72
väli[fi], 47
väliaika[fi], 47
väliakseli[fi], 21
väliholkki[fi], 88
välitiukkuus[fi], 92
välitys[fi], 40

väljennin[fi], 69
väljentää[fi], 69
väljyys[fi], 6
välys[fi], 6, 41
vändskär[sv], 46
väritaso[fi], 87
värma[sv], 43
värmebehandling[sv], 43
värmebeständig[sv], 43
värmevärde[sv], 43
væske[nb], 50
väsyminen[fi], 33
vätska[sv], 50
växel[sv], 40
växellåda[sv], 40
växelström[sv], 3
växla[sv], 13
väännin[fi], 12, 25
vääntiö[fi], 12
vääntiölaikka[fi], 49
vääntömomentti[fi], 91
våglängd[sv], 96
vågrät[sv], 44

W

wagon, 96
washer, 26, 96
washing machine, 74
washing nozzle, 96
waste rock, 77
water cooled, 96
water level, 50
water pump pliers, 96
water separator, 96
wave length, 96
wear, 96
wear resistance, 96
wearing steel, 96
weave, 97
web, 71
wedge, 97
weld, 97
weld penetration, 98
weld pool, 56
welder, 97
welding, 97
welding apparatus, 97
welding burner, 97
welding electrode, 97
welding helmet, 97
welding machine, 97
welding seam, 97
welding torch, 97
welding transformer, 97
welding wire, 98
wheel, 98
wheel nut wrench, 98
wheel puller, 67
whetstone, 98
whetting, 98
wick, 98
wick lubrication, 98
width across flats, 98
winch, 98
wind, 98
winding, 98
wing nut, 98
wire, 11, 98
wire drawing, 99
wire[sv], 98
wiring diagram, 99
wood saw, 99
wood-screw, 99
working environment, 99
working machine, 99
working margin, 99
working stroke, 65
workpiece, 99
works, 53
workshop, 99
worm, 99
worm gear, 99
worm wheel, 99
wrench, 81

Y

yield point, 99
yksitoiminen sylinteri[fi], 79
yleisjyrsinkone[fi], 94
yleiskulmamitta[fi], 76
yleiskulmamittain[fi], 94
yleisnivel[fi], 94
yleispihdit[fi], 17
ylempi myötöraja[fi], 99
ylikuormitus[fi], 60
ylipaine[fi], 60
ympyräura[fi], 49
yritys[fi], 53
yta[sv], 86
ytbehandling[sv], 87
ytnormal[sv], 87
yträhet[sv], 87
ytterring[nb], 59
ytterring[sv], 59
yxa[sv], 5

Z

zero point, 99
zink-plating, 100

Æ

ädelgas[sv], 57
äes[fi], 43
ämne[sv], 53
ändplanfres[sv], 77
ässja[sv], 38
äänenvaimennin[fi], 78, 79
äänitorvi[fi], 45

Ø

ögleskruv[sv], 32
øks[nb], 5
øksehammar[nn], 5
øksehammer[nb], 5
öljy[fi], 58
öljyallas[fi], 58
öljykannu[fi], 58
öljykylpy[fi], 58
öppen ringnyckel[sv], 36
øreklokker[nb], 28
øreplugg[nb], 28
ørepropp[nb], 28
öronpropp[sv], 28
överbelastning[sv], 60
överföring[sv], 92
överlappning[sv], 60
övertryck[sv], 60
övre sträckgräns[sv], 99
øyebolt[nb], 32
øyre-[nn], 28
øyreklokker[nn], 28

Å

ånga[sv], 85
ångmaskin[sv], 85
åpen ringnøkkel[nb], 36

www.ingramcontent.com/pod-product-compliance
Ingram Content Group UK Ltd.
Pitfield, Milton Keynes, MK11 3LW, UK
UKHW022000270726
14060UKWH00003B/617

9 788293 828068